Entrepreneurship Development Series Volume 03

AGRICULTURAL PROJECT MANAGEMENT

NIPA GENX ELECTRONIC RESOURCES & SOLUTIONS P. LTD.

New Delhi-110 034

Entrepreneurship Development Series Volume 03

AGRICULTURAL PROJECT MANAGEMENT

Lakshmi Dhar Hatai
M.SC. (Ag), Ph. D. (BHU)
Associate Professor (Agricultural Economics)
Department of Social Science
College of Horticulture and Forestry
Central Agricultural University
Pasighat, Arunachal Pradesh-791102

NIPA GENX ELECTRONIC RESOURCES & SOLUTIONS P. LTD.
New Delhi-110 034

**NIPA GENX ELECTRONIC
RESOURCES & SOLUTIONS P. LTD.**

101,103, Vikas Surya Plaza, CU Block
L.S.C.Market, Pitam Pura, New Delhi-110 034
Ph : +91 11 27341616, 27341717, 27341718
E-mail:newindiapublishingagency@gmail.com
www: www.nipabooks.com

For customer assistance, please contact
Phone: + 91-11-27 34 17 17
Fax: + 91-11-27 34 16 16
E-Mail: feedbacks@nipabooks.com

ISBN: 978-93-94490-31-4

Composed and Designed by NIPA.

Preface

Project plays a pivotal role in fostering and sustaining the tempo of rural development. Project is the essential element for agricultural development. Agricultural project is an investment activity where we spend capital resources to create a productive asset for realizing benefits over time. It consists of a series of organized and supervised activities with start and finish dates that are carried out to meet specific requirements, such as time and cost. Agriculture is the emerging growing sector contributing towards improving farm incomes, enhancing food and nutritional security, reducing rural poverty and accelerating the overall economic growth of our country. There is considerable emphasis on enhancing agricultural growth through effective project management in the areas of technological changes, area expansion, cropping intensity and value addition by attracting private investment in agro-processing sector. Project management is the discipline of planning, organizing and managing resources to bring about the successful completion of specific project goals and objectives.

The book is divided into ten chapters that dealing with all the relevant topics in the field agricultural project and its' management. The book represents the synopsis of the project management in agriculture. All these chapters have been presented in a logical sequence, simple, lucid style with full emphasis on basic concepts of the subjects. This book has signified and provides basic knowledge about project identification, formulation, analysis, implementation, monitoring and evaluation that will be helpful for effective project evaluation and its management in agricultural development. Chapters listed in the book deals with project cycle, project design, market and demand analysis, analysis of stakeholders and appraisal techniques etc. in an agricultural project. Contents of this book would go a long way in guiding the students, managers, academicians, policy makers, economists, researchers and entrepreneurs in improving their knowledge level on agricultural project management. The book encourages the reader to treasure and periodically refer this book as it is a rare collection of what is found generally in the literature. I trust, the readers will reap maximum benefits from this book and I am confident that the book would get wider acceptance by the students and teachers dealing with agricultural project and it's management. I am especially indebted to the authors whose works are cited throughout the book and acknowledge to individuals, institutions and publishers in which the articles is originally appeared. I dedicate this book to the stakeholders related to agricultural project and it's management.

Lakshmi Dhar Hatai

Contents

1

Project and Project Management

Agricultural project is an investment activity in which financial resources are expended to create capital assets that produce benefits over an extended period of time. Projects are the building blocks of investment plan. The whole complex of activities in the undertaking that uses resources to gain benefits constitutes the agricultural project. Project is a specific activity with a specific starting point and a specific ending point intended to accomplish a specific objective. It is something which is measurable both in its major costs and returns. It will have some geographic location or at least a rather clearly understood area of geographic concentration. It will have a specific clientele group which it is intended to reach. It will have a relatively well defined time sequence of investment and production activities.

Meaning and Concept of Project

- Projects are the essential elements for agricultural and socio-economic development.
- Project is derived from the latin word "Projectum"—(meaning)—"to throw something forward".
- A project is a temporary endeavor, having a defined beginning and end, undertaken to meet unique goals and objectives, usually to bring about beneficial change or added value.
- Project is a scheduled set of activities aimed towards the creation of particular asset as per planned specifications, with a view to generate wealth – as estimated – for coming years.
- A project is a non-routine, non-repetitive undertaking.
- The relationships in a project setting are dynamic, temporary and flexible.
- A project requires coordination of the efforts of persons drawn from different functional areas and contributions of external agencies.
- A project is a combination of inputs managed in a certain way to achieve one or more desired outputs and ultimately one or more desired impacts.

Inputs include people (staff, partners, personnel, stakeholders, government officials), equipment (vehicles, machines, computers) supplies and communications (paper, phone, e-mail etc), travel (to bring other inputs together) and training.

- Project may be an experiment or a study. It may be strategic or applied. It may be simple or large.
- Project is a bundle of activities designed to get the desired results.

P-Problem

R-Rhythm

O-Objectivity

J-Justified

E-Effort

C-Creativity

T-Time bound

Project is a single-shot set of activities having a definite beginning and ending points. The activities of a project are performed in a particular order as they have precedence relationships. The key concept of a project is that it is a one-time occurrence, an occurrence that will not be repeated daily, weekly or monthly converting organisational resources into goods or services.

Construction of a house, establishing a plant or factory, arranging a birthday party or picnic, preparing for a competitive examination, municipal construction of new tennis court at a park and initial production of a new product are all beginning and ending points and each has a series of activities with precedence relationships.

Project is an organised programme of activities carried out to reach a defined goal, often of a non-recurring nature with a specified terminal point. It is a time-bound package of scheduled and assembled activities dedicated to the attainment of a specific objective of successful completion of a work, on time and within the allotted budget.

Various scholars, practitioners and management scientists have dealt with the concept of 'project' in their own way. According to **F.L. Harrison,** "A project can be defined as a non-routine, non-respective, one-off undertaking, normally with discrete time, financial and technical performance goals."

In the words of **Little and Mirrlees,** "We mean by a project any scheme, or part of a scheme, for investing resources which can reasonably be analysed and evaluated as an independent unit."

According to the **Encyclopaedia of Management,** project is "an organised unit dedicated to the attainment of a goal-the successful completion of a development project in time within budget, in conformance with pre-determined programme specification."

Project is defined in the **Dictionary of Management** as "an investment carried out according to a plan in order to achieve a define objective, within a certain time and which will cease when the objective is achieved."

To quote **Sinha and Sinha**, "A project is not a mere action or an activity or an attempt towards a particular aim; it is rather an integrated effort, including multifarious actions and activities, towards that aim."

The report of the **Economic Commission for Asia and Far East** (ECAFE) describe that "Project is a smallest unit of investment activity to be considered in the case of programming. It will, as a rule, be a technically coherent undertaking which has to be carried out technically speaking, independently of other projects." Similarly, according to the **Food and Agriculture Organisation** (FAO), a project is "an activity sufficiently self-contained to permit financial and commercial analysis in most cases projects represent the expenditure of capital funds by pre-existing entities which want to prove their operation. It is convenient to divide all projects into two broad classes: (i) those that are revenue producing and self-financing (i.e., commercial type projects), and (ii) social projects."

All the above definitions thus suggest that a project is an action-oriented undertaking. It is a one-shot, time-tested, goal-directed, major undertaking requiring the commitment of varied skills and resources. **According to the Project Management Institute, USA, *"A project is an undertaking with a define objective by which completion is identified. In practice, most projects depend on finite or limited resources by which objectives are to be accomplished."***

A project is also defined as a non-routine, non-repetitive, one-off undertaking normally with-discrete time, financial and technical performance goals. It can be considered to be any series of activities and task that:

- Have a specific objective to be achieved within certain specifications;
- Have defined start and end dates;
- Have funding limits (if applicable);
- Consume resources (time, money, equipment etc.) and
- Generate wealth (tangible or intangible).

Finally, in short, we can define project as, **"Project is a non-routine one-off undertaking for future with specific performance goal".**

Project Definition

- Project is an investment activity where we spend capital resources to create a productive asset for realizing benefits over time.
- Project is an activity on which we spend money in expectation of returns, which lends itself to planning, financing and implementation as a unit.
- Project also refers to specific activity, with specific starting point and specific end points to achieve a specific objective. It should be measurable in costs and returns. It must have priorities for area development and reach specific clientele groups.
- Project is an investment activity meant for providing the returns for specific clientele group for specific activity, specific objective and specific area development. It must be facilitating analysis in planning, financing implementation, monitoring controlling and evaluation.

Characteristics of Project

A project has several characteristics. The characteristic features of a project are briefly described below:

- A project has a mission or a set of objectives. Once the mission is achieved the project is treated as completed.
- A project has to terminate at some time or the other; it cannot continue forever. The set of objectives indicate the terminal stage of the project.
- While the numbers of participants in a project are several, the project is one single entity and its responsibility is assigned to one single agency.
- A project calls for team-work the members of the team may come from different organizational units, different disciplines, and even from different geographic regions.
- A project has a life cycle represented by growth, maturity and decay. A project has a learning component.
- A project is unique and no two projects are similar, even though the plants set up are identical. The organizations, the infrastructure, the location and the people make the project unique.
- Change is a natural phenomenon with every project throughout its life span.
- Some changes may not have any major impact, but some others may change the very nature of the project.
- The happenings during the life cycle of a project are not fully known at any stage. As time passes, the details are finalized successively. For example, more details are known about the project at the erection stage than at the detailed engineering stage.

- A project is always customer-specific. The requirements and constrains within which a project must be executed are stipulated by the customer.
- A project is a complex set of things. Projects vary in terms of technology, equipment and materials, machinery and people, work ethics and organizational culture.
- A substantial portion of the work in a project is done by sub-contracting. The greater the complexity of a project, the greater will be the extent of work performed by subcontractors.
- Any project is exposed to risk and uncertainty and the extent of these two depend upon has the project moves through the various stages in its life span.
- A well defined project has lesser risk and uncertainty, whereas an ill-defined project faces greater degree of risk and uncertainty.

The main characteristics of a project include the following features

i) **Mission or set of objectives:** A project has a mission or a set of objectives to be achieved within a distinct time, cost and technical performance goals. Once the mission is achieved, the project is treated as completed.

ii) **Ownership:** Every project has an owner. Who, in the private sector can be individual or a company; in the public sector, a government or government undertaking; and in a joint sector organisation, the ownership of a project can be represented by a partnership of public and private sector.

iii) **Terminal stage:** A project cannot continue forever. It has to terminate at some time or the other. The set of objective of a project indicate the terminal stage of the project.

iv) **Team work:** Every project is planned, managed and controlled by an assigned team- the project team-to achieve the objectives as per specifications. The members of the project team may come from different organisational units, different disciplines and even from different geographical regions.

v) **Uniqueness:** A project is unique and no two projects are similar, even though the inputs, processing and results of two projects are identical. The nature of an organisation, available infrastructure, location of the project and the people associated with a project make a project unique.

vi) **Risk and uncertainty:** Every project has risk and uncertainty associated with it and there cannot be a project without any risk and uncertainty. The risk is perceived to be variability of actual returns from the estimated returns and uncertainty about future leads to variations in returns. The degree of risk and uncertainty also very during the life-cycle of a project. An ill-defined project has a high degree of risk and uncertainty in comparison of a well-defined project.

vii) **Sub-contracting:** A substantial portion of each and every project is performed by sub-contracting. The greater the complexity of a project, the greater will be the extent of work performed by sub-contractors.

viii) **Customer-specific:** A project is always customer-specific. The requirements and constraints within which a project is executed are generally stipulated by the customer or are planned considering the beneficiary customer.

ix) **Life-cycle:** Every project has a start and a terminal point- a characteristic of a life-cycle represented by growth, maturity and decay. The organisation of a project changes as it passes through this cycle. The activities starting from the conception stage, mounting up to the peak during maturity or implementation, comes to the zero level on completion and delivery of the project.

x) **Management of change:** Change is a natural phenomenon with every project throughout its life-cycle. Project varies during its life-span in terms of technology, equipments, materials, machinery, people, work ethics and organisational culture. Some changes in these variables may not have any major impact but some others may change the very nature of a project. Therefore, the inter-relationship and management of change between these variables is required for successful completion of a project.

xi) **Unity in diversity:** Every project has the characteristic feature of unity in diversity. A project is a complex set of thousands of different activities which remain inter-related with each other. Although these activities differ in terms of materials, machines, equipments, technology, people, ethics, work culture etc. Yet these are essentially inter-related during different stages of a project.

Meaning of Project Management

Due to these reasons, project management requires a different form of organization, sharper tools of planning and control, and improved means of coping with human problems.

- Management is the process of designing and maintaining an environment in which individuals, working together in groups, efficiently accomplish selected aims.
- Management is carrying out the managerial functions of planning, organizing, staffing, leading, and controlling.

- It applies at all levels of project.
- It applies to any kind of project.
- It is concerned with productivity; this implies effectiveness and efficiency.

Management a very widely used term with varying perceptions has been defined in different ways as:

a) Getting the things done
b) The ability of a person to accomplish the particular task activity. Despite hardships
c) Using available resources effectively
d) The ability of a person in getting results by using his techniques and skills for optimizing benefits.

Projects are cutting edges of development. Projects are means for increasing the output from the given resources. Evaluation of project needs projecting the future trend of output; a sales expected cost, returns, flow of funds etc. (cash flow).

- Project management is the discipline of planning, organizing and managing resources to bring about the successful completion of specific project goals and objectives.
- Project management is the planning, organizing, directing and controlling of organizational resources for a relatively short and / orlong term objectives that have been establish to achieve specific goals and objectives.
- Project management is the discipline of organizing and managing resources in such a way that these resources deliver all the work required to complete a project within defined scope, time, and cost constraints.
- Project management can be defined as the discipline of applying specific processes and principles to initiate, plan, execute and manage the way that new initiatives or changes are implemented within an organization.

Successful project management can be defined as having achieved the project objective:

i) Within the time frame
ii) Within cost
iii) At the desired performance
iv) While utilizing assigned resources effectively and efficiently.

Project is a system involving co-ordination of a number of separate department entities throughout the organization and which must be completed within prescribed schedules and time constraints. Managing a project involves application of scientific tools like operation research, quantitative techniques, qualitative techniques, planning, implementing, monitoring and coordinating all the activities or tasks to produce desirable outputs in accordance with the predetermined objectives.(According to Project Management Institute, USA)

Key components of project management

- Time – the intended duration of the work
- Cost – the budget allocated for the work
- Scope – what innovations or changes will be delivered by the project
- Quality – the standard of the outcome of the project.

Project management stages

Although there are different project management methodologies and approaches, most projects follow these stages:

- Initiating the project: The project manager defines what the project will achieve and realize, working with the project sponsor and stakeholders to agree deliverables.
- Planning: The project manager records all the tasks and assigns deadlines for each as well as stating the relationships and dependencies between each activity.
- Execution: The project manager builds the project team and also collects and allocates the resources and budget available to specific tasks.
- Monitoring: The project manager oversees the progress of project work and updates the project plans to reflect actual performance.
- Closing: The project manager ensures the outputs delivered by the project are accepted by the business and closes down the project team.

Projects are cutting edge of development. Therefore, we need to identify national agricultural development objectives for selection of priority areas for developmental plans require good projects and good projects require sound planning. Both are interdependent. Project planning seeks to ensure optimization of scarce resources for balanced growth of economy. It should facilitate an analysis in planning, financing, implementation, monitoring, controlling and evaluation.

Characteristics of Agricultural Projects

1. Project is made up of many sub-projects / investments. Ex: Sericulture, palm oil industry, ice factories for fisheries project, construction of 500 dug wells (each well a sub-project).

2. Project increases capital intensity of farm enterprise
3. Investment costs vary according to natural conditions, market conditions and processing conditions
4. The quantum of incremental benefits (income) depends on benefitting area and the level of benefits depends on stage of development farms
5. The level of benefits may, however fluctuate from year to year due to weather conditions.

Types of projects

Agricultural Projects

1. **Water Resources Development Project-** This project includes irrigation, ground water project, a project for land reclamation, drainage project, salinity prevent and flood control project.
2. **Agricultural Credit Project -** These are also called (on-lending project). This project provides credit to the farmer for farm investment for increasing agricultural production raising their standard of living and the economy as a whole. – Commercial Banks- RRB, PACs, Co-operatives.
3. **Agricultural Development Projects** – These projects aim at improving farm economy of individual and regional development.
4. **Agro-Industries and Commercial Development Projects** – Projects of input supply, services to farming, projects related to processing, storage, market development etc.

Development Projects

A development project has the objective of improving something – to make life better for a group or another in same way. Ex: Government spends much of its budget on development projects which range from building of roads and schools to the provision of drinking water, training to teachers, doctors and agricultural development functionaries.

Research Projects

A research project seeks to overcome a constraint or solve a problem. It will yield an output that will be knowledge or a new technology, but it will not usually directly yield developmental objectives. However the results of a research project very often are the basis for development projects.

Project Classification

Projects have been classified in various ways by different authorities. Little and Mirrelees divide the project into two broad categories, viz., quantifiable projects and non-quantifiable projects. The planning Commission has accepted the sector criteria for classification of projects. Projects can also be classified on the basis of techno-economic characteristics. All India financial institutions classify the projects on the basis of the nature of the project and its life cycle. The project classifications are explained below.

1. Quantifiable and non-quantifiable projects

Quantifiable are those in which a plausible quantitative assessment of benefits can be made. Non-quantifiable projects are those where such as an assessment is not possible. Projects concerned with industrial development, power generation, mineral development are forming part of quantifiable projects. The non-quantifiable projects category comprise health, education and defence.

2. Sectoral Projects

According to the Indian Planning Commission, a project may fall in the following sectors:

a) Agriculture and Allied Sector

b) Irrigation and Power Sector

c) Industry and Mining Sector

d) Transport and Communication Sector

e) Social Service Sector

f) Miscellaneous Sector

3. Techno-Economic Projects

Techno-economic projects classification includes factors intensity-oriented classification, causation-oriented classification and magnitude-oriented classification. These three groupings are narrated as under:

a) **Factor intensity oriented classification:** The factor intensity is used as based for classification of projects such as capital-intensive or labour-intensive which depends upon the large scale investment in plant and machinery or human resources.

b) **Causation oriented classification:** The causation-oriented project are determine base, on its cause namely demand base or raw material-base project. Th3e non-availability of certain goods or services and consequent demand for such goods or service or the availability of certain raw material, skills or other inputs is the dominant reason for starting the project.

c) **Magnitude oriented classification:** The size of investments forms the basis for magnitude-oriented projects. Projects may thus be classified base on its investment such as large-scale, medium-scale and small-scale projects.

Techno-economic characteristics-based classification is useful in facilitating the process of feasibility appraisal. United Nations and its specialised agencies use the International Standard Industrial Classification of all economic activities (ISIC) in collection and compilation of economic data. Since this classification covers the entire field of human economic endeavour, if forms a useful basis for classification of projects. Economic activities are under this classification grouped into ten divisions, which are sub-divided into ninety sub-divisions. The divisions are

- Division 0_________Agriculture, Forestry, Hunting and Fishing
- Division 1_________Mining and quarrying
- Division 2 and 3_________Manufacturing
- Division 4_________Construction
- Division 5_________Electricity, Gas, Water and Sanitary Services
- Division 6_________Commerce
- Division 7_________Transport, Storage and Communications
- Division 8_________Services
- Division 9_________Activity not adequately described

4. Financial Institutions Classification

All India and State Financial Institution classify the projects according to their age and experience and the purpose for which the project is being taken up. They are as follows:

i) New project

ii) Expansion projects

iii) Modernisation projects

iv) Diversification projects

The projects listed above are generally profit-oriented and the services oriented projects are classified as under:

i) Welfare Projects

ii) Services Projects

iii) Research and Development Projects
iv) Educational Projects

World Bank has recognized six important aspects in the project preparation.

i) Technical
ii) Administrative
iii) Organization
iv) Commercial
v) Financial
vi) Economic aspects

i) Technical

All the technical aspect assists of the project must be thoroughly study under technical analysis goods and services are required for project execution needs for assessment. For awareness of the lending agency – Bank, NGO's financial Institution regarding the technology to be used (Capital Intensive Technology)

- Labour Intensive Technology
- Latest / Existing technology.

ii) Administrative

Coverage, managerial aspects project staffs, extension personal, credit agency and farmers. (Beneficiary will be study)

iii) Organizational

Deals with relationship of project administration and the governments or training arrangements, disbursement of payment (wages) etc.

iv) Commercial Aspects

Regarding commercial of the project arrangements of supply of inputs materials, services needed for the project, marketing of output etc. are to be assets.

v) Financial

Regarding financial aspects, the item which form under this category are sources of funds, cause of finds, repayment etc. The estimated cost based on technical aspects, estimated sales based on commercial analysis and are probable profits from the operation of the projects are to be properly evaluated.

vi) Economic Aspects

On the other land, economic analysis concentrates in determine project contribution to the development of the economy as a whole and justifying the huge of scarce resources, proper identification of cost and benefit is an important aspect in the economic analysis of project.

Need for Project Management

Most of the organization whether public / private, produces a product or provide a service or both have some basic premises for project management. All of these organization work to perform so that their goods / services are provided at the high time at minimum cost (more profit) at the service of the client consumer.

Project management has very wide range of availability. Its present use only scratches the surface is equally useful in small / large organization and small / large reach project.

Small project can also benefit from the concern attention of a project manager with the advantages of better budgetary and civil control & better allocation of resources.The principle of project managers are applicable to every organizer dealing with. In the past, project mgt has been widely used in the defence department the aerospace Industry and major construction project. Project mgt is getting acceptance in product development public works, Non-profit organization, financial institute in all sector of government.

What does Project Management does for you?

Once top management have decided there so considered the use of project management. They must review what it can do for them. Are the benefit of both of implementing both and very upsetting to their organization? What are the benefits? What do they by implementing project management? Project management will centralized responsibility for the following key item in one individual.

- Budgeting cost control
- Schedules
- Resource allocation
- Technical quality
- Client, customer, Public Relation

Budgeting cost control

Budgeting is the part of the planning function and also serve as a control mechanism giving the basis from which actual performance can be conferred.

Major expenses and corrected, control accounts are identified and established the budget to be reviewed and used as a programme for plan expenditure. When the budget is combined with a scheduled. It serves as a programme of physical completion and provide a most important tools of project management.

Schedules

A scheduled however must be a living document. Easily understood, previously committed by its participants and readily adjusted to replay rent project condition. The scheduled should show the interrelationship of one group plant achievement with other groups. If a project manager does is to think out each of the key steps of design and logical activities sequence. The significance progress milestone towards completing the project.

Resource allocation

Each project requires some resources in the procurement of materials equipment and services in orderly manner. Without that orderly organization it is highly probable that plan cost / and / or schedule will not be made due to delay of delivery of key item. A material procurement for cash is an important-document to manage the resources and it can identified they step in the procurement process from department of specifications. Through the Bid cycle Vendor Prints ending with delivery dates for any development defect in material & equipment desires.

Technical quality

Once a quality control system has designed & documented a designed review processed is in order. This should being together export. 2nd level procedure to review as a committee a pre design segment of work with to manufactured and construction. A "good hold and witness" control plan must be developed with its specification. Start up an acceptance should be outline well in advance Methodology & special measuring device must be adequate.

Client Customers, Public Relations

Once of the most important aspect of the project manager job is the responsibility as a representative of top management. The project must recognize a clear fully plan of day to day contact at working levels. The project will eventually become by pass when the customer / clients need information. It is quietly likely that top management will then find project manager usefulness indeed. Public relation are not always concerned to the project manager, however if the project is sensitive to the public interest because of environment marketing / safety factor than project manager is deeply involved.

2

Pre Feasibility Study and Project Cycle

Pre-feasibility study

A well planned projects requires pre-feasibility study to explore the opportunity to find out problems, avenues of co-ordination etc. A government being a regulation spells out priotize for national and regional economy and by the frameworks for resource allocation.

A survey is required to assets the needs of the farming community and area and priotization of needs and problems. Survey also ensures the availability of raw material & skills and manpower of the agricultural environment. Publication of government financial institution, consultation organization, NGO's research institution helpful in getting desire information. An analysis of the economic and social trend will be very helpful in identifying and projecting the needs of intervention and project.

Pre-feasibility analysis seeks to determine whether the project is prima facie worthwhile to justify a feasibility study/techno-economic feasibility study and what aspect of the project are critical to its viability. When a project opportunity study is undertaken in respect of an investment opportunity, the pre-feasibility analysis can be dispensed. Similarly, when a sector or resource opportunity study contains sufficient project data to either proceed to the feasibility stage or to determine its discontinuance, pre-feasibility analysis can also be by-passed. However, this analysis is conducted when the economies of the project are doubtful unless a certain aspect of the analysis has been investigated in depth to determine the viability of the project, either by a detailed market study or by some other functional study. While conducting pre-feasibility analysis, short-cuts may be used to determine minor components of investment outlay and production cost but not to determine major cost components. These components must be estimated for the project as a part of the pre-feasibility analysis.

Objectives of Pre-feasibility Study

The main objectives of pre-feasibility study are to determine the following:

- Whether the investment opportunity is so promising that an investment decision can be taken on the basis of information elaborated at the pre-feasibility stage?
- Whether the project concept justify a detail analysis by a techno—economic feasibility study?
- Whether the information is adequate enough to decide that the project is not either viable proposition or attractive enough for a particular investor or investor group?
- Whether any aspects of the project is critical to its feasibility and require in-depth investigation through functional or support studies such as market surveys, laboratory test, pilot plan tests etc.

Any proposal for a project generally passes through the three stages of scrutiny and clearance-prefeasibility, project feasibility and detailed project report. The structure/outline of the information and estimates in which they are prepared and submitted is the same for all the three stages of scrutiny. The difference is only in the refinement or level of accuracy to be achieved in the data, information and figures as the project formulation or project design advances. Refinement at a later stage can be supported by more detailed estimates, graphs, worksheet, questions and the like because a more clear report will elicit a faster clearance. Supporting documents, explanatory notes, sketches, diagrams etc. may be given with the pre-feasibility report to facilitate appraisal. Accordingly, it is necessary even at the pre-feasibility stage to ensure that the project is viable from the following angles.

- Market demand for end-products of the project.
- Material and other inputs.
- Plant location and site.
- Project investment costs.
- Project engineering-technologies, equipments and engineering works.
- Plant implementation schedule.
- Financial and economic evaluation including commercial profitability.
- Statutory clearance.

On the basis of pre-feasibility analysis a report is prepared, known as pre-feasibility report (PFR) which elicits the preliminary sanction or first stage clearance by the Board of Directors and/or the Government. Generally, it is not expected from a PFR to contain precise details and accurate figures but an attempt should be made to give the best available information. When an aspect, either favourable or adverse, calls for special attention in making a decision, that aspect should be highlighted in sufficient details e.g.., the cost of scientific farm technology/processing technology to be acquired.

Pre-feasibility study is required to

i) To determine whether projects offer a promising investment opportunity.

ii) To determine whether the project will be able to reap social benefits.

iii) To determine whether any aspects of the projects case critical requiring in depth investigation by the way of survey (PRA & others).

iv) To study the projects of other line agencies like govt. NGO's about the activities, operational area of project management.

v) To study the required and availability of man-power resources and project cost.

If the pre-feasibility study indicates that the project is worthwhile proposition, a detail pre-feasibility study is taken out.

The management of social development project involves apply of different managerial in all the stages of project cycle. This skill are apply in decision making, supervision, budgeting and reporting.

Project Cycle

The project cycle illustrates the set of actions, design, planning, implementation, monitoring, evaluation, reporting and learning.

The project cycle shown a circle because of insight and learning from project evaluation, inform the design of new projects learning after and learning before.

Phases / Stages of Project Cycle

1. Conception / Identification
2. Formulation / Preparation of project
3. Appraisal/ Analysis
4. Implementation
5. Monitoring
6. Evaluation

Project is considered as a cycle because each phase not only grow out of the preceding one. But leads into the subsequent phase and it is a self renewing cycle so that new projects come out of the old one in a continuous manner.

1. Concept / Identification of project

In several agricultural projects cost are easier to identify than benefits because the expenditure pattern is easily visualized. Various type of cost involved in the project are:

i) **Project Cost:** This includes the value of resources in maintaining and operating of the project.

ii) **Associated Cost:** Cost that incurred to produce immediate product and services of the project for use / sales.

iii) **Primary Cost / Direct Cost:** This includes cost incurred in construction maintenance and execution of the project.

iv) **Secondary / Indirect Cost:** Value of goods and services incurred in providing indirect benefits from the projects. e.g. School, hostel.

v) **Real Cost and Nominal Cost:** Cost at current market price are. Nominal cost whereas if cost are deflated by general price index, these are called as real cost.

vi) **Social Cost:** These are technological externalities and technological spill-over accrued to the society due to the presence of projects. e.g. Water and Soil Pollution, Health hazards etc.

Tangible benefits: Incremental income due to existence of project as. Its obtained from the following changes.

→ Improvement in chopping pattern including high value crops.

→ An increase in productivity of crops.

→ Increase in the intensity of cropping.

→ Reduction in the cost of cultivation of crops.

→ Large scale economics due to specialization.

Intangible benefit: These include better income distribution, national integration and better standard of living etc.

In the process of identification / conception. The important sub stages are

a) Preliminary study

b) Pre feasibility study

c) Project report

In these stages we assess whether the project proposed on the grounds of prima-facie is feasible and the objectives of the project achieved. On this ground, the preliminary study should embody the investment proposals, benefits extended from the projects and method of implementation. Assessment of the demand for the project's products, technical feasibility of the project, import and export requirements, marketing aspects, investment prospects, etc., should be exhaustively covered by the feasibility studies, including the analysis of sensitivity. Some of the sources through which the projects identified are:

- Agricultural and allied programmes proposed in the plans of the country as well as states
- Areas identified as potential for further development through Governmental surveys
- Special developmental programmes like RKVY, PMFBY, PMKSY, PMKISAN
- Irrigation projects which offer scope for development through forward and backward linkages.
- New projects emerging out of existing projects, etc.

2. Formulation/ Preparation

The location of the project & project site must be based on technical analysis technical feasibility of the project. The location of the project depend upon available feasible resources, marketing concept, marketing facility alternative investment project, alternative experience, farmer objective, technical skill motivation, demand for product. Technical analysis must take into consider all aspect of technology to be used in the project. In the account to be used in goods & services. Assessment of suitability and adequacy of natural resources based on scientific study is also essential. Alternatives to the resource used are to be considered in the formulation of the project.

Due consider is to be given

a) Technical
b) Financial
c) Commercial
d) Managerial
e) Organization
f) Social
g) Economic aspects

Identification of the missing links in the infrastructure particularly in relation to communication system, market storage facilities are important.

a) **Technical aspect:** The issues which needs technical examination are thoroughly analyzed. For eg: For agricultural project related to pre production, production and post production.

b) **Financial aspect:** We have to find out the source of raising financial assistance and term and for getting finance. The implementing agencies should be in a position to financial assistant anticipated results through farm planning and farm budgeting.

c) **Commercial aspect:** This aspect focus on the estimation of effective demand, availability of input supply & arrangement for output market.

d) **Managerial aspect:** Once identification process is over we have to find out the managerial competence of the beneficiaries. The managerial skill can be serpended if necessary technical skills are imparted by the extension agencies we have the responsibility of technology transfer.

e) **Organizational aspect:** Organization refers to the process of looking the priority in prepare the organization of hierarchy of the implementing agency the availability of staff at various status, demarcation of authority & linking of authority are included under this aspect.

f) **Social aspect:** Custom, culture and tradition of the beneficiary are considered the other relevant implication of the probable changes in the living standard. Material welfare, consumption, income distribution and impact fall under this aspect.

g) **Economic aspects:** Here, we have to examine the benefits in which the project is going to contribute in term of utilization of scarce resources of the nation.

3. Appraisal / Analysis

Appraisal should occur before the implementation of the project. It is done independently by specialists. In the appraisal stage it is important to know whether the project is technically feasible according to the data available. The technical data for the assessing the feasibility of the project should be consistent, with the information available in the office of the – authority or elsewhere. Managerial aspects play a key role in the project appraisal. New skill and information gained by the beneficiary or farmer in the project area including adoption of new technology.

4. Implementation

This is most crucial phase of the project cycle. The secret of the successful implementation depend upon the extend of realism put into the plans drawn before hand. It is often project implementation can be divided into three different periods

1. Investment periods.
2. Development periods
3. Full production periods

Investment periods may range from few months to few years depending upon the nature of assets to be required. Assets proposed should be superior quality.

Development period too consume time. Implementing agency should make all effort to reduce the gestation period as per the plan that included in begining.

Full production period is the time during which the beneficiaries start reaping the benefits of the project. Implementing agency should ensure that the project beneficiary do continue to receive the benefits during the entire life of the project.

5. Monitoring

Monitoring is the timely collection and analysis of data on the progress of a project with objective of identified constraint which impede successful implementation. This is highly desirable particularly when the project fail, to be completed as per the time schedule / in the process of attaining the set of goals. This imperative to get the feedback on the problem faced so that effective measure can be taken up to overcome the deficiency which hamper the speedy implementation. Monitoring has to been done continuously to offset various shortcomings or problem that crop up from time to time. in various aspects of implementation.

6. Evaluation

It is the last stage of project cycle. It is not confined to completed project. Evaluation can be done several times during the life of a project. Evaluation process is important to see how for the objective set out in the project are achieved. Deficiency or failures to achieve the objectives may be analysed and addressed proper solutions to such failures.

Evaluation completed in 3 phase

1. Pre-project Evaluation / Mid courses Evaluation
2. Concurrent Evaluation
3. Ex-post Evaluation/ End Evaluation

Mid course evaluation

In the first phase- Evaluation is attempted before any change occur in the existing situation. This is primarily mean to assess economic feasibility of the projects, since it done at the very beginning (Pre-project Evaluation)

Concurrent Evaluation

Such type of evaluation to take out when the project is execution for identifying and analyzing the pitfalls in the project.

Ex-post Evaluation / End Evaluation

Evaluation is a resorted to particularly when the project is completed in all its phase in order to assess the achievements of end / objective set out by the projects. Such evaluation is called Expost evaluation / End evaluation.

Evaluation is done by the agency other than the implementing one, viz. Financial bank / sponsoring agency or government.

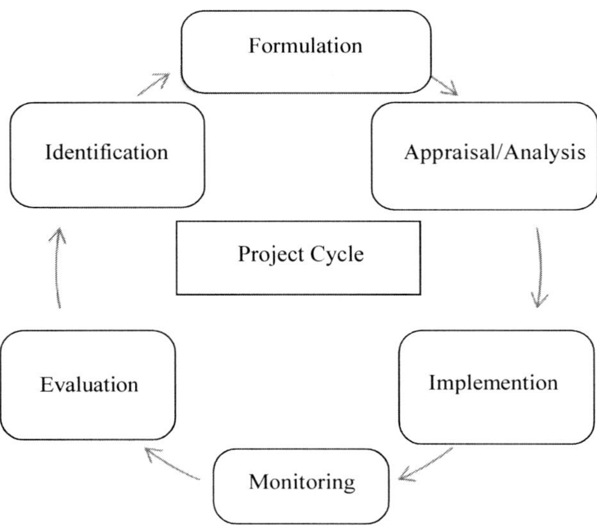

3

Project Design

Meaning of Project Design

Designing is a creative activity. Think of fashion designers or interior decorators, who design clothes or decorate homes. Think of the people who design sports cars or bikes, or the interiors of fancy hotels or corporate offices. They take simple things like fabric, or steel, or plastic and using colors create things that are attractive, useful and a pleasure to live with. Designing a project is equally creative - selecting inputs and managing them in a certain way to achieve objectives.

Project Design is the formal framework which establishes in advance the extend work plan the means of ascertaining progress and assumption on which the project is based. The premise of design is simply that thinking a project through will improve its chance of achieving something and that such effort used to be invested before the project start.

A good agricultural project document we answer the following question

1. What is the agri project expected to accomplished if the project successfully completed on time.
2. Why is the agricultural project is being undertaken?
3. How is the agri project to be implemented?
4. Who is primarily responsible for the agri project implementation?
5. Who are the intended beneficiaries, the targeted group we are to be expected to benefits projects.
6. Within what period of time the project to be carried out & the objective to be attained.
7. What resources are necessary to achieve the objectives?
8. What external factor are necessary for agri project success?

A will design project is usually characterized by project document which logical and complete the format outline provides the planner with a place to put each item of information and permits verification. The process usually begins with the identification of problem followed by formula of objective, which spell out the decision, action / change in target group / area which is expected to occur as a direct result of undertaking.

Project Design components

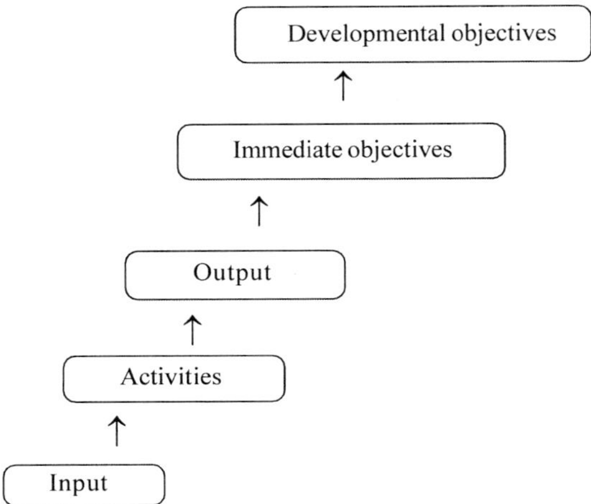

A Project document prepared along the following outline provides a solid & foundation for subsequent implementation. It will also serve as evaluate basis for evaluation since prior management about what the manage can and should be able to achieve is the only fair standard against which result can be measured.

Sections

1. Background and content
2. Target group (Intended beneficiary)
3. Objective

1. **Background:** This section of the project document describing the problem / situation which has been observed and which seems to come for action by donor. The statement of background which should be lost of serve to sane the stage, to frame the problem for which a solution is short. It should also described past effort, especially by the donor to address the problem and the result
2. **Target group:** Whose the project being conducted ? What population group / groups are expected to be better off either directly or in the of as a result

of the project. Many thing such as livelihood, poverty Alleviation, livestock management, agricultural production etc are conceived with broad target group when it comes to design them however the specific target group often are not clearly identified as a result, it may be discovered at the end of the project. Therefore sound project required the identification of the target group and major which will be to access the project impact on it. While description such as the rural poor under employed workers may be appropriate for the estimate of policy. They are clearly inadequate for individual budgeting document. A prices prescription of the target group will help ensure that the benefit of the given project to indeed flow to them.

If specific identification is not feasible the computation of target group may be narrow do acc. to one / more of the following criteria.

1. Geographical
2. Field activity of farmer with certain crop or income landless labour artisan etc.
3. Lack of access or to certain service (health, education etc).
4. Nutritional status, assets measures (land holding shelters, livestock etc).

3. **Objective:** What specific effect is the project expected to achieve within its lifetime and what impact is it so have on the target group?

   a) **Formulative Objectives:** A project idea arises when a problem / situation is observed which seems to invite donor actions. A rough formulation of the objective can often be arrived at inverting the problem.

      I. **Problem** – The government is unable to carry out main power requirement study to outside assistant. This impacts the development of suitable manpower policies more than 1/3rd of rural labour population is under employed / semi employed.

      II. **Inversion** – Govt. is able to easy out manpower recognizes study without assistant. With the assistant of expert NGO's and other effective policies are developed as a result viewer people we under employed.

   b) **Means and ends:** An objective is a simple expression of desire ends. In formulating an objective we should not confused end such term as to promote, to coordinate to guide etc did not means rather than ends should not used when starting objectives in order to avoid confusion. Those projects elements which are causes and those which are effect.

   c) **Multiple Objectives:** In general, project should be formulation in terms of single development objective often denotes ambiguity or presence of means rather than ends. When there are multiple objective at either level such objective compliment each other / complete with each other only on the carefully circumstances. Can do or more completing objective peacefully exist.

eg : Enhance the income of farmers by adopting the modern cultivation practices.

- Create a better understanding of the impact of population land ownership through establishment of research sentries.

Inputs

What money, personnel materials, services etc are to be provided by the donor, funding agencies and /or the govt?

Major input by all key parties should convey a realistic picture of what is to be provided by the donor /common man. Input should normally listed by source (funding agencies) and other donors and should shows – type (Expert, Equipment, fellowship etc).

- Number / amount
- Land / duration of assignment
- Costs
- Qualification or field of specialization
- By which to be provided
- Try to fix a realistic data
- Purpose of which provided

Activities

What activities need to be undertaken in order to produce the desire output? Activities are the action, the research, the task to be carried out by the project staff. This usually described what project magnitude does the means by which inputs are transformed into Output. For each output there will normally be one / more activities. State the activities and verify that.

i) Activities are so stated that there implementation can be verify in terms of quantity and time.

ii) Activities are stated in terms of action being undertaken. All key activities are necessary for achieving the output.

iii) There are no activities listed where effect can't be traced upward to the output level.

Output / Result

What kind and how many in to be produced with the inputs provided and activities undertaken in order that the immediate objective might be achieved.

Outputs are the product of completed activities. A such output differ substantially from the object which is the effect of the project hopes to achieve.

State the Output then verify that

i) Outputs are so stated that their realization can be verify in terms of quality of time.

ii) Output are stated as the result of activities.

iii) All key inputs necessary for achieving the immediate objective are listed.

iv) Each Output can be trace upward immediate objective level.

Immediate Objective

What specific effect is the project to achieve within its lifetime or directly life after. If the project is completed successfully what improvement or changes are expected in the group organization or area toward which the project is directed?

The immediate objective is the situation that is expected to prevent at the end of project. This level of the project design obviously is a crucial importance since it will determine the magnitude of input, activities and plan outputs. As well as the process or approach to be employed. Once the immediate objective has been formulated verify that the state may says what effect will have been achieved if the project is completed successfully and on scheduled. The project is stated in term of single immediate objective not a collection of sequencely link target.

The immediate objected is stated in terms of end or not of means There is reasonable expectation that the achieve of immediate objective will make meaningful contribution towards the attainment of the development objective.

Development Objective

What is the result for the broader category objective toward which the effort of the objective is directed why the project is undertaken?

A single project can't normally be expected to bridge about the achievement of the development objective by remove basic constrain or contributing or more necessary elements. Once the development objective has been formulated verify that it reflects the result for the project.

i) Comprises & single objective or a group of comfortable objective.

ii) It shows stated that progress work can be verify.

Project Proposal -Meaning

A plan or suggestion, especially a written one, put forward for consideration or discussion by others (donors). Proposal has a format:

- It should propose a systematic investigation targeted to a specialist slot on the R&D continuum.
- The scale of application and its domain must be clearly defined.
- The scientific research methodology should be detailed in the project proposal.
- The role and the aims of research project should be described in terms of the utilization of the new knowledge towards sustainable production system.

A winning research project proposal is the one that typically meets the donor's objectives and needs of the society at large and finally,

- A project proposal that will win a grant would be innovative; it would explore new areas of research and would try to discover new techniques/knowledge. The project results when adopted on a large scale would make the world a safer place to live and would aid in sustainable development.

Project Designing Key Concepts

Goal

A goal is an over-arching target or reason for something. Ex: Identifying poverty alleviation and protection of the environment are very large goals.If the goal is more specific, it becomes much easier to measure. All goals have something to do with making something better, improving something or preventing something bad from happening.

Objective

An objective is simpler and more specific than goal. Goal may be improving health; for this goal, the objective, which contributes to the goal, may be to do daily exercise. So objectives contribute to goals, but alone do not achieve goals. Just as a project is part of a program, an objective is part of a goal.

Project output

If objectives are achieved, that is the project output or result. In research projects the output may be tangible, like a new crop variety or a new type of farm equipment or a new vaccine or invisible, like weight reduction due to daily exercise. However, in all cases, the output is something "new ", in the sense that it was not there before.

Project's impact

Project impact is the expected benefit to the beneficiaries. The output alone cannot achieve the goal of the project. But it will continue towards the goal, by contributing to the project's impact. If daily exercises are done, health will soon improve. Of course, daily exercises alone will not make a person healthy, if he / she is eating too much or drinking too much, or smoking too much. But daily exercises will help to a certain extent.

Principles of good project design

Understanding the following principles of project development can help you to become a good project designer:

1. Project design is both an art and a science

The art is using imagination to envisage the future, to fantasize about what activities can help to improve the lives of others, and to write persuasively and well. The science is to employ logic in the design of project. Everyone can help these artistic and scientific skills. As with other endeavors, practice makes perfect.

2. Project design is a group exercise

Project design is not a solitary activity. Always a team will do better than an individual. A team of people from different backgrounds will usually do better than a team of people with the same training and the same attitudes. In the business of agricultural research, it's been found that a team or a consortium that includes partners and beneficiaries–people from all sorts of disciplines, ages and backgrounds, designing together in a true spirit of collaboration yields the most successful (and most likely to be funded) projects of all.

3. Project design takes time

Developing a relatively simple and small project can make the main designer in anywhere from 50-150 hours of quality time, spread over several months. Complex projects involving multiple partners and sometimes multiple donors will take much longer time.

Small – budget projects often take almost as long to design as larger ones. For this reason, there may be many advantages to thinking big. However, big grants that need to be approved at a donor headquarters may take longer to get approved.

In addition, to the time for thinking and writing, you need to add the time spent for waiting. Even when a donor likes a project, there is more to be done. Few projects are funded as submitted. Most require revision, based on donor's comments.Patience and persistence are essential project writing skills.

Be prepared for failure

One should not expect to win every proposal submitted. If competing for funds, it is clear that not all submissions can be funded.

Partners are important

Almost all research projects nowadays involve collaboration among different stakeholders and colleague groups. There are some special points about partnerships to remember when studying to be a good project designer.

Donors (like other investors) prefer to share and spread risks. They like to fund projects that have attached support from others. Further, donors prefer projects with multiple partners. There is no doubt that projects designed and implemented by partners (consortium) are the pattern of the day and also future.

Recognize the "towards holistic thinking"

For example several of the UN organizations were created in an effort to integrate approaches or see things holistically. A more recent example is the establishment of global entities to coordinate the sectors concerned with water – including sanitation and supply, irrigation, environment and industry.

When reviewing proposals today, donors will look favorably at projects that adopt and integrated approach that involves all the participants and is thus likely to contribute to significant impacts in the shortest possible time.

Demonstrating impact is essential

In this competitive scene, project outputs (new technologies, more people trained) alone can no longer sell projects. Today donors are looking for projects that show positive impact on problems like poverty, malnutrition, ill health, population migration and environmental degradation. What is more, donors are looking for impacts that can be demonstrated quickly, ideally during the project itself.

Impact assessment is harder than measuring the number of people trained, miles of road built, seedlings distributed and tons produced. There are lots of difficulties in assessing the impact. What about differences those cannot be measure? Will the benefits last? Who should take the credit? But in spite of these difficulties, organizations need to think about impact assessment, and how to measure and demonstrate impact if the project is successful in attracting external donor funds.

Packaging is what it is all about?

In packing research problems into projects, designers need to consider the issues already mentioned, the importance of partners and impact, the need for patience and persistence, and need to understand the self-interest of the donors they are writing to. All this makes up the art of "selling" the project, and that is what happens when an attempt is made to write a proposal.

Meaning of a Concept note

A concept note (or concept paper, as some call it) is a summary version of a proposal.

- A concept note for internal approval may be as short as one or two pages.
- A concept note for discussion with partners will be a little longer.
- A concept note for submission to a donor is ideally between three and seven pages long.

There are three stages in developing a project.

1. Internal approval.
2. Obtaining inputs from partners.
3. Submitting a project to a donor.

A concept note is used for stage 3 i.e., while submitting a project to a donor. Preparation of full proposal takes a long time. Designer will not want to spend this time unless he is certain that his/her proposal will be read and given due consideration. Donors are busy people, so to make a donor really interested, the first approach should be through the submission of a short note.

A concept note has to be prepared if

- Submitting a sole proposal, and / or
- To find out if a donor might be interested, or
- Ideas are at a preliminary stage.

Should not start with a concept note if

- Responding to a request for a proposal, or
- Applying for a grant under a competitive grants program, or
- A donor has said that a full proposal would be welcome.

Concept note:Order of preparation

1. Objectives
2. Inputs
3. Activities and duration
4. Outputs
5. Beneficiaries and impacts
6. Project managements
7. Draft budget
8. Background
   a) The problem and why it is urgent
   b) What has already been done

Steps in Project preparation

Step: 1. Objectives

Step: 2. Inputs

Step: 3. Activities and Duration

Step: 4. Outputs (what will be achieved at the end of the project?)

Step: 5. Beneficiaries and Impacts

Step: 6. Project management

Step: 7. Estimated Budget

Step: 8. Background Material

Step: 1.Objectives

The objectives are the single most important part of project design. They tell the reader what the designer want to do.Project objectives should

a) Correspond to a core problem,

b) Define the strategy to overcome the problem, and

c) Contribute to the achievement of higher-level development goals.

When formulating objectives, keep in mind that objectives should be **SMART!!**

- **Specific**
- **Measurable**
- **Achievements**
- **Realistic**
- **Time bound**

Each objective should specify the quantity of achievement (e.g., number of beneficiaries, number of new varieties developed), and the quality (e.g., poor rural women/farmers, high-yielding varieties). Objectives should also include an indication of when the objective will be achieved (e.g., in January 2019 three years after the start of the project). The following three words may be kept in mind.

- **Quality**
- **Quantity**
- **Time**

Step: 2. Inputs

The inputs needed to implement the project (i.e. to achieve objectives) may include:

i) People (project leader and partner's staff line)

ii) Travel costs (TA, DA and other incidentals)

iii) Vehicles

iv) Equipment (tools, scientific, office)

v) Supplies (stationary, chemicals, seed, fertilizer, etc.)

vi) Services (telephone, fax, email, transportation, dispatch, electricity, etc.)

vii) Facilities (offices, library, training center, demonstration plots, etc.)

Some inputs may come from others, e.g., farmer groups, individual farm families, other Non-Government Organizations, donor groups, government agencies, etc. remember that all partners will also have travel, supplies, services and other input requirements.

Step: 3. Activities and Duration

Briefly describe the project leader and his/her partners plan to achieve the project objectives.

Tips

- **Be brief and clear.**
- **Be positive**
- **Do not use "we".**

Step: 4. Outputs (what will be achieved at the end of the project?)

The outputs of the project should be directly related to the project objectives. Out puts may include:

- Events, such as workshops or harvests.
- Intangible things, like decisions.
- Tangible things, like new building and new facilities.
- Information, perhaps in the form of publication or videos.

It is worth spending time with colleagues, partners and friends brainstorming all the possible spin-off outputs, as well as those directly related to the objectives.

Step: 5. Beneficiaries and Impacts

Brainstorm this section with the design team or other colleagues. Think of all the possible groups who may benefit from project activities and as many different benefits as may occur.

Impact is what the donor is "buying". In making promises about the impact of a project, you need to:

- Describe the expected benefits, how many of them can be expected, and when and where they will occur.
- Reasons for the expected benefits to accrue to a given group – if necessary, state the assumptions.
- Consider whether to suggest that the project will have either an impact assessment component or will be assessed by a separate impact measurement project.

Possible beneficiary groups

- Poor individuals (age? sex? location?)
- Farm families (including dependents)
- Refugees
- Poor urban consumers
- Other population groups

Show impact in terms of goals, such as:

- Poverty alleviation
- Food security
- Preserving the environment
- Improved nutrition and health

Develop impact checklist

Will the project result in

- More education for the poor?
- Better health for poor families?
- Gender-specific or age-specific impact?
- Enhanced community participation?
- New use of indigenous Knowledge?
- Inputs for improved decision- making?
- New jobs created?
- Other economic benefits? which sectors?
- Improved child nutrition?
- Other human benefits?.

Step: 6. Project management

The best objectives in the world can only achieve the desires outputs and impacts, if the projects can be effectively managed. Project design needs to include a plan covering the roles and responsibilities of the various people who will manage the project.

Step: 7. Estimated Budget

Budgeting is as important for development of organizations as for multinational corporations. Even top-quality proposals will not get funded if their cost estimates are unrealistic, overly greedy, or fill gaps that will cause future delays and frustrations. Budget preparation skills are an essential tool for all who seek funds.

In a concept note, an estimate of the project cost has to be given by a rough costing of the main project inputs, generously rounded up. An allowance for the budget of possible partners and indirect costs are also have to be included.

Be sure to include and label all project costs. It is very important for all parties to understand the true and full project costs, and to avoid hidden subsidies.

Step: 8. Background Material

In the concept note, organize background material in two sections.

1. Under **"The problem and why it is Urgent,"** discuss the project in terms of goals: ex: poverty alleviation, food security, preservation of the environment, and nutrition and health.

2. Under **"What Has Already Been Done,"** Do not focus only on own activities. Donors will want to acknowledge the contribution others have made and are still making- some may be organizations that are supporting and some may be proposed partners. (If this is a follow – on project or second phase, describe the outcomes of the earlier work in detail.)

Selecting a good title

Titles needs to be catchy, informative and distinctive. Try to use a two-part title. The first part should be short, snappy and catchy; the second part can be more serious and informative. Never use words like "studies on........", "investigation on........", "Survey of ...." "Effect of.........", "standardization of ...." , etc.,in the titles.

Example

- Empowering Rural Women for Sustainable Family Improvements

  Source : ANGRAU, Hyderabad.
- Biodiversity of Flesh Fungi: An Alternative Source of Food for NE Region

  Source : IARI, New Delhi.

Project formulating definition

Introduction

Project formulation involves creating a project plan outlining the activities, tasks, dependencies and timeframes. The steps involved in project initiation are:

1. **Generation of project ideas**
2. **Monitoring of environment**
3. **Corporate appraisal**
4. **Suggestions for exploration of project ideas**
5. **Preliminary screening**
6. **Project rating index**

Generation of ideas

Most of the project ideas involve combining existing fields of technology or offering alternatives of present products or services. The typical route may be described as follows:

- A person with specialized technical knowledge or marketing expertise feels that he can offer a product or service which can cater to a presently unmet need.

- He can serve a market where demand exceeds supply or effectively compete with similar product or services because of certain favorable features like better quality or lower prices.
- His ideas are endorsed by his associates who encourage him and even show willingness to collaborate with him on the proposal.
- Finally he receives support from financial institutions and banks that approve his project and show readiness to finance it.

Stimulating the flow of ideas: To stimulate the flow of ideas, the following are helpful:

SWOT Analysis : SWOT is an acronym for strengths, weaknesses, opportunities, and threats. SWOT analysis represents a conscious, purposeful and systematic effort by an organization to identify opportunities that can be profitably exploited by it. Periodic SWOT analysis facilitates the generation of ideas.

Clear articulation of objectives : The operational objectives of an organisation may be one or more of the following:

- Cost reduction
- Productivity improvement
- Increase in capacity utilization
- Improvement in contribution margin

A clear articulation and prioritization of objectives helps in channelizing the efforts of employees and stimulate them to think more imaginatively.

Fostering a conducive climate: To tap the creativity of people and to harness their entrepreneurial urges, a conducive organizational climate has to be fostered. Ex: motivating employees through awards, incentives, promotions to think and work more creatively.

Monitoring the environment

Basically a promising investment idea enables a firm (or entrepreneur) to exploit opportunities in the environment by drawing on its competitive strengths. Hence, the firm must systematically monitor the environment and assess its competitive abilities. For purposes of monitoring, the business environment may be divided into six broad sectors.

- The key sectors of the environment are as follows:
- **Economic sector**
- State of economy
    - o Overall rate of growth

  - Growth rate of primary, secondary and tertiary sectors
  - Cyclical fluctuations
  - Inflation rate
  - Linkage with the world economy
  - Trade surplus / deficits
  - Balance of payment situation
- **Government sector**
- Industrial policy
  - Government programmes and projects
  - Tax frame work
  - Subsidies, incentives, and concessions
  - Import and export policies
  - Financing norms
  - Lending condition of financial institutions and commercial banks
- **Technological sector**
- Emergence of new technologies
  - Access to technical know-how, foreign as well as indigenous
  - Receptiveness on the part of industry
- **Socio-demographic sector**
  - Population trends
  - Age shifts in population
  - Income distribution
  - Educational profile
  - Employment of women
  - Attitudes toward consumption and investment
- **Competition sector**
  - Number of firms in the industry and the market share of the top few(four or five)
  - Degree of homogeneity and differentiation among products
  - Entry barriers
  - Comparison with substitutes in terms of quality, price, appeal, and functional performance
  - Marketing policies and practices

- **Supplier sector**
- Availability and cost of raw materials and sub-assemblies
    - Availability and cost of energy
    - Availability and cost of money

Corporate appraisal

A realistic appraisal of corporate strengths and weaknesses is essential for identifying investment opportunities which can be profitably exploited. The broad areas of corporate appraisal and the important aspects to be considered under them are as follows:

1. Marketing and Distribution

- Market image
- Product line
- Market share
- Distribution share
- Customer loyalty
- Marketing and distribution costs

2. Production and Operations

- Condition and capacity of plant and machinery
- Availability of raw materials, sub-assemblies and power
- Degree of vertical integration
- Locational advantage
- Cost structure

3. Research and development

- Research capabilities of the firm
- Track record of new product developments
- Laboratories and testing facilities
- Coordination between research and operations

4. Corporate Resources and Personnel

- Corporate image
- Clout with governmental and regulatory agencies

- Dynamism of top management
- Competence and commitment of employees
- State of industrial relations

5. Finance and Accounting

- Financial leverage and borrowing capacity
- Cost of capital
- Tax situation
- Relations with shareholders and creditors
- Accounting and control system
- Cash flows and liquidity

Suggestions for exploration of project ideas

Good project ideas – the key to success – are indefinable. So a wide variety of sources should be tapped to identify them.

1. Analyze the Performance of Existing Industries

- It is essential to study the existing industries in terms of their profitability and capacity utilization.
- This analysis of profitability and break – even level of various industries indicates promising investment opportunities which are profitable and relatively risk free.
- A study of capacity utilization of various industries provides information about the potential for further information.

2. Examine the Inputs and Outputs of Various Industries

An analysis of the inputs required for various industries may throw up project ideas.

- Existing opportunities for procuring materials or supplies including transportation costs are to be considered.
- Several firms produce internally some components / parts which can be supplied at a lower cost by a single manufacturer.
- Similarly, a study of the outputs of the existing industries may reveal opportunities for adding value through further processing of main outputs, by – products, as well as waste products. Remember that one person's trash can be another person's treasure.

3. Review Imports and Exports

An analysis of import statistics for a period of five to seven years is helpful in understanding the trend of imports of various goods. An examination of statistics is useful in learning about the export possibilities of various products.

4. Study Plan Outlays and Governmental Guidelines

A very valuable source of information to estimate the scope for further investment is the **guidelines to industries** published annually by the department of industrial development, government of India. This publication provides information about the structure and location, production performance, licensed and installed capacity, exports, and future scope of various industries.

5. Suggestions of Financial Institutions and Developmental Agencies

In a bid to promote development of industries in their respective states, state financial corporations, state industrial development corporations, and other developmental bodies conduct studies, prepare feasibility reports, and offer suggestions to potential entrepreneurs. The suggestions of these agencies are helpful in identifying promising projects.

6. Investigate Local Materials and Resources

A search for project ideas may begin with an investigation into local resources and skills. Various ways of adding value to locally available materials may be examined. Similarly, the skills of local artisans may suggest products that may be profitably produced and marketed.

The national council of applied economic research (NCAER) and other bodies publish surveys of various regions showing the potential of industrial development in various regions. These surveys assess the resources (human and material), infrastructural facilities, and markets for various products.

7. Analyze Economic and Social Trends

Changing economic conditions and consumer preferences provide new business opportunities. For example, a greater awareness of the value of time is dawning on the public. Hence, the demand for time – saving products like prepared food items, ovens, and powered vehicles has been increasing. Another change that can be seen is the increasing desire for leisure and recreational activities. This has caused a growth in the market for recreational products and services.

8. Study New Technological Developments

There is a large network of research laboratories in India under the umbrella of the council of scientific and industrial research and other bodies. New products or new processes and technologies for existing products developed by research laboratories like CFTRI(Center for Food Technology and Research Institute) may be examined for profitable commercialization.

9. Draw Clues from Consumption Abroad

Entrepreneurs willing to take higher risks may identify projects for the manufacture of products or supply of services which are new to the country but extensively used abroad. Automatic vending machines, entertainment parks, pre- fabricated houses, and fast food restaurants are examples of projects belonging to this category.

10. Explore the Possibility of Reviving Sick Units

A significant proportion of sick units can be nursed back to health by sound management, infusion of further capital, and provision of complementary inputs. Hence, there is a shorter gestation period because one does not have to begin from scratch.

11. Identify Unfulfilled Psychological Needs

For well – established, multi-brand product groups like bathing soaps, detergents, cosmetics, and tooth pastes, the question to be asked is not whether there is an opportunity to manufacture something to satisfy and actual physical need but whether there are certain psychological needs of consumers which are presently unfulfilled. To find out whether such an opportunity exists, the technique of spectrum analysis is useful. This analysis is done in the following manner:

- Important factors influencing brand choice are identified.
- Existing brands in the market are positioned on a continuum in respect of the factors identified.
- Gaps which exist in relation to consumer psychological needs are identified.

12. Attend Trade Fairs

National and international trade fairs provide an excellent opportunity to get to know about new products and developments.

13. Stimulating Creativity for Generating New Product Ideas

New product ideas may be generated by thinking along the following lines: modification, rearrangement, reversal, magnification, reduction, substitution, adaptation, combination. Identification of investment opportunity may be influenced by the chance factor also.

Preliminary screening

The following aspects may be considered for preliminary screening.

- Compatibility with the promoter: The idea must be compatible with the interest, personality, and resources of the entrepreneur.
- Consistency with governmental priorities: The project idea must be feasible to the national goals and governmental regulatory framework
- Availability of inputs: The resources and inputs required for the project must be reasonably assured.
- Adequacy of market: To judge the adequacy of the market the following factors have to be examined.
    - Total present domestic market
    - Competitors and their market shares
    - Export markets
    - Quality – price profile of the product vis-à-vis competitive products
    - Sales and distribution system
    - Projected increase in consumption
    - Barriers to the entry of new units
    - Economic, social, and demographic trends favorable to increased consumption
    - Patent protection

Reasonableness of cost

The cost structure of the proposed project must enable it to realize an acceptable profit with a competitive price. The following should be examined in this regard.

- Costs of material inputs
- Labor costs
- Factory overheads
- General administration expenses
- Selling and distribution costs
- Service costs
- Economies of scale

Acceptability of risk level

The desirability of a project is critically dependent on the risk characterizing it and the following factors should be considered:

- Vulnerability to business cycles
- Technological changes
- Competition from substitutes
- Competition from imports
- Governmental control over price and distribution

Project rating index: The steps involved in project rating index are as follows.

- Identify the factors relevant for project rating. (Flow chart)
- Assign rates to these factors.
- Rate the project proposal on various factors, using a suitable rating scale. Ex: 5 point or 7 point scale.
- For each factor multiply the factor rating with the factor weight to get the factor score.
- Add all the factor scores to get the overall project rating index.

Agricultural Project Frame Work / Outline of project

Part – I

1. Title of the agricultural project
2. Location
3. Background Information
4. Intended beneficiaries / Target group.
5. Goal (the higher order objective to which the project contribute).
6. Purpose (the designing part of the project).
7. Development Institution.
8. Project activities
9. Duration of the project
10. Project participation (Project design, management) monitoring and evaluation).
11. Project Management (how the project will be managed, responsibilities of implementation, monitoring and evaluation.
12. Project Sustainability (economic, social and ecology).

(How will be activities with be sustained after the subject to be continue).

13. Budget (the calculation of the budget needs to be based on input needed to carry out the activities and the prices and cost currently applicable in the project area.

Head	1st year	2nd year and so on
1. Salary		
2. Equipment		
3. Travel		
4. Raw material for receiving		
5. Recurring expenses		
6. Any other		

14. Justified of budget (item wise)
15. Contribution from NGO
16. Contribution from community
17. Yield or output from project
18. Impact of agricultural project

Part – II (Institutional framework).

1. Name of the organization
2. Address of the organization
3. Legal status
4. Society Legislation Act / Any other 80G Income Tax Act: Name and Designation of chief functionaries.
5. Brief Bio-data of the project functionaries
6. Present activities of the organization
7. A / C of Bank details
8. Any other specific information.

Part – III (Enclosure)

1. Registration certificate
2. Certified by laws
3. Audited statement of account
4. Annual reports
5. Any other relevant document

4

Market and Demand Analysis

Concept and Meaning of Market

The word market is derived from the latin word "marcatus" which means "merchandise or trade or a place where business is conducted ". The term market has been widely used in various means, such as:

- A place or a building where commodities are bought and sold (e.g. Super market)
- Potential buyers and sellers of a product (e.g. cotton market, wheat market)
- Potential buyers and sellers of a country or region (e.g. Indian market, Asian market)
- An organization which provides facilities for exchange of commodities (e.g. Bombay Stock Exchange)
- A phase or a course of commercial activity (e.g. a dull market or bright market).

In India, the markets are otherwise known as: Hats, Painths, Shandies and Bazar.

Definitions of Market

1. Market is the sphere within which price determining forces operate.
2. Market is the area within which the forces of demand and supply coverage to establish a single price.
3. Market includes any place where persons assemble for the sale or purchase of commodities intended for satisfying human wants.
4. The term market means not a particular market place in which things are bought and sold but the whole of any region in which buyers and sellers are free to interact with one another that the prices of the same goods tend equally, easily and quickly.
5. Market means a social institution, which performs activities and provides facilities for exchanging commodities between buyers and sellers.

6. Economically interpreted the term market refers, not to a place but to a commodity or commodities and sellers are in free interact with one another.

Components of market

The following conditions must be required for existence of a market.

a) Existence of a good or commodity for transactions (physical existence is, however not necessary)
b) Existence of buyers and sellers
c) Business relationship or intercourse between buyers and sellers
d) Demarcation of area such as place, region, country or the whole world.

Definitions of Marketing

1. Marketing is the performance of business activities that direct the flow of goods and services from the producers to consumers.
2. Marketing contains a series of activities involved in moving the goods from the point of production to the point of consumption. It includes all the activities involved in the creation of time, place, form and possession utility.
3. Marketing is the economic process by which goods and services are exchanged between producers and the consumers and their values determined in terms of money prices.
4. Marketing is creating value for customers
5. Marketing is the delivery of a standard of living to society.
6. Marketing is designed to bring about desired exchanges with target audiences for the purpose of mutual gain.
7. Marketing activities are concerned with the demand stimulating and demand fulfilling efforts of the enterprises.
8. Marketing is a total system of interacting business activities designed to plan, promote and distribute need-satisfying products and services to existing and potential consumers.
9. Marketing starts with the identification of a specific need on the part of consumer and ends with the satisfaction of that need. The consumer is found both at the beginning and the end of the marketing process.
10. Marketing originates with the recognition of a need on the part of a consumer and terminates with the satisfaction of that need by the delivery of a usable product at the right time, at the right place and at an acceptable price.
11. Marketing is the process of planning and executing the conception, pricing, promotion, and distribution of ideas, goods and services to create exchanges

that satisfy individual and organizational goals (According to American Marketing Association).

Agricultural Marketing

According to Acharya and Agarwal, Agricultural marketing refers to the study of all the activities, agencies and policies involved in the procurement of farm inputs by farmers and the movement of agricultural products from the point of production to the point of consumption. This agricultural marketing system is a link between the farm and non-farm sectors.

According to National Commission on Agriculture (NCA) in XII report, Agricultural marketing is a process which starts with a decision to produce a saleable farm commodity, and it involves all the aspects of market structure and system, both functional and institutional, based on technical and economic considerations, and includes pre- and post harvest operations, assembling, grading, storage, transportation and distribution.

Characteristics of Agricultural Commodities

i) Perishability of agricultural commodities

ii) Seasonality of agricultural production

iii) Bulkiness of agricultural products

iv) Large variation in quality of agricultural products

v) Uncertain and irregular supply of agricultural products

vi) Small size of holdings and scattered production

vii) Inadequate processing facilities for agricultural commodities

Classifications of markets

The markets are classified on the basis of following aspects:

1. Location or Place of operation
2. Area or coverage
3. Time span
4. Volume of transactions
5. Nature of transactions
6. Number of commodities
7. Degree of competition
8. Nature of commodities
9. Stage of marketing

10. Extent of public intervention
11. Type of population served
12. Market functionaries and accrual of marketing margins

1. Based on Location

i) Village market: This market is located in a small village, where major transactions take place among the buyers and sellers of that village.

ii) Primary wholesale markets: These markets are located in big towns near agricultural commodities production point.Usually, Transactions take place between the farmers and traders.

iii) Secondary whole sale markets: These markets are located generally in district headquarters or important trade centres or near railway junctions where, transactions take place between the village traders and wholesalers.

iv) Terminal Markets: The market where the produce is either finally disposed off to the consumers or processors or assembled for export. Merchants are well-organised and use modern methods of marketing. Commodity exchanges exist in these markets which provide facilities for forward trading. These markets are located near the metropolitan cities.

v) Sea board markets: These markets are mainly for import and /or export purpose and located in sea shores.

2. Based on Area/ Coverage

i) Local /village markets: A market in which the buying and selling activities are confined among the buyers and sellers drawn from the same village or nearby villages. The market deals with perishable commodities in small lots. (e.g. village market)

ii) Regional markets: A market in which buyers and sellers for a commodity are drawn from a larger area than the local markets. Regional markets in India usually exist for foodgrains.

iii) National markets: A market in which buyers and sellers are at the national level. These are found for durable goods like jute and tea.

iv) World Markets: A market in which the buyers and sellers are drawn from the whole world. These are the biggest markets from the area point of view and world wide demand and supply such as coffee, tea, machinery, gold, silver etc.

3. Based on Time span

i) Short-period markets: The markets which are held only for a few hours, products are a highly perishable in nature (e.g. fresh vegetables, fruits, fish, milk etc.)

ii) Long-period markets: These markets are held for a longer period than the short period markets. Commodities are less perishable and can be stored for some period. (e.g.foodgrains, Oilseeds etc.).

iii) Secular markets: These markets are permanent in nature. The commodities traded are durable in nature and stored for many years. (e.g. machinery and manufactured goods)

4. Based on Volume of transaction

i) Wholesale markets: A wholesale market is one in which the commodity are bought and sold in large lots or in bulk. Usually, the transactions take place mainly between traders.

ii) Retail markets: A retail market is one in which commodities are bought by and sold to the consumers as per their requirements. Transaction takes place between retailers and consumers.

5. Based on nature of transaction

i) Spot or cash markets: A market in which goods are exchanged for money immediately after sale.

ii) Forward markets: A market in which the purchase and sale of a commodity takes place at time 't' but the exchange of the commodity takes place on specified date in the future, i.e. time 't+1'.

6. Based on number of commodities in which transactions take place

i) General markets: A market in which all types of commodities (foodgrains, oilseeds, fibre crops, FMCGs, clothing and apparel) etc. are bought and sold (large no. of commodities)

ii) Specialized markets: A market in which transactions takeplace only in one or two commodities. For every group of commodities separate markets exist. (e.g.foodgrain markets, vegetable markets, wool market, cotton market)

7. Based on degree of competition

i) Perfect market: A perfect market in which the following conditions holds good:

a) There is large number of buyers and sellers

b) All the buyers and sellers in the market have perfect knowledge of demand, supply and prices.

c) Prices at any one time are uniform over a geographical area, plus or minus the cost of getting supplies from surplus to deficit areas.

d) The prices are uniform at any one place over periods of time, plus or minus the cost of storage from one period to another.

e) The prices of different forms of a product are uniform plus or minus the cost of converting the products from one form to another. Otherwise, the various characteristics of perfect markets are:

   - Large number of buyers and sellers
   - Homogeneous products
   - Free entry and exit of firms
   - No government regulation
   - Perfect mobility of resources
   - Perfect knowledge

ii) Imperfect markets: The markets in which the conditions of perfect competition are lacking are called imperfect markets. The following situations, each based on the degree of imperfection, may be identified:

a) Monopoly market: (mono = single, poly = seller). It is a market situation in which there is only one seller of a commodity. He exercises sole control over the quality or price of the commodity. The price of a commodity is generally higher than in other markets. (e.g. Indian farmers operate in a monopoly market when purchasing electricity for irrigation, Indian post)

b) Monopsony: It is a market situation when there is only one buyer of a product.

c) Duopoly: It is a market situation which has only two sellers of a commodity.

d) Duopsony: It is a market situation which has only two buyers of a commodity.

e) Oligopoly: A market in which there are more than two but still a few sellers of a commodity.

f) Oligopsony: A market having a few (more than two) buyers.

g) Bilateral monopoly: It is a market structure in which a single seller faces a single buyer.

h) Monopolistic competition: It is a situation where large number of sellers deal in heterogeneous and differentiated form of a commodity. Different prices prevail for the same basic products.

8. Based on nature of commodities

i) Commodity markets: A market which deals in goods and raw materials such as wheat, cotton, fertilizers, seeds etc. There is existence of necessary facilities for proper transaction and efficient selling of particular commodity in large quantity.

ii) Capital market: The market in which bonds, shares and securities are bought and sold. e.g. Money markets and share markets.

9. Based on stage of marketing

i) Producing markets: These markets which mainly assemble the commodity for further distribution to other markets. These markets are located in producing areas.

ii) Consuming markets: Markets which collect the produce for final disposal to the consuming population.

10. Based on extent of public intervention

i) Regulated Markets: The markets in which business is done in accordance with the rules and regulation framed by the statutory market organization and represent different situations involved in markets, the marketing costs in such markets are standardized and marketing practices are regulated.

ii) Unregulated markets: Business is conducted without any set of rules and regulations. Traders frame the rules for the conduct of business and run the market. These markets suffer from many ills, ranging from unstandardized charges for marketing functions to imperfections in the determination of prices.

11. Based on Type of population served

i) Urban Market: This market serves mainly the population residing in an urban area. The nature and quantity of demand for agricultural product is originating from the urban population.

ii) Rural Market: This market is concerned with the flow of goods and services are taking place from urban to rural and vice versa and within the rural area itself. There is inflow of products into rural markets for production or consumption and also outflow of products to urban areas.

(Rural to urban flow: Rice, Wheat, Sugar, Tobacco, Cotton and vegetables etc.

Urban to rural flow: Fast Moving Consumers Goods (FMCGs), Textiles and consumer durables)

12. Based on Market functionaries and accrual of marketing margin

The markets are also classified based on who are the market functionaries and to whom the marketing margin accrue. There is considerable increase in the producer and consumer cooperatives and other organization performing marketing of various agricultural commodities. Based on this the markets may be:

i) Farmer market

ii) Cooperative markets and

iii) General markets.

Marketing Strategy

Marketing strategy is an integral part of marketing planning. In modern day business terms, strategy refers to the matching of the activities of an organization to the environment in which it operates and to its own resource capabilities.

Definition: Marketing strategy is the complete and unbeatable plan, designed especially for attaining the marketing objectives of the firm/business unit. The marketing objectives indicate what the firm wants to achieve, the marketing strategy provides the design for achieving them.

e.g. marketing objective of a business unit in the next year to achieve sales revenue of Rs. 1,000 crore and net profit will be 15 per cent on sales revenue. Then it is the job of marketing strategy to indicate how and wherefrom this sale and profit will come, which product lines/products /brands will accomplish this task and how?

Specifies the position the unit will seek in its industry and how it will compete there in?

1. What position does the unit seek in its industry?
2. What market segments to serve and what product offers to make?
3. The growth path: market penetration, market development or product development
4. Who are my competitors? Whom to compete, whom to avoid?
5. On what differentiation strength to compete? (product superiority, brand power, distribution strength and better service)
6. On what competitive advantages will the fight be based?
7. What are the resources?

Formulating the marketing strategy

Basically, formulation of marketing strategy consists of three main tasks:

- Selecting the target market
- Positioning the offers
- Assembling the marketing mix

This implies that the essence of the marketing strategy of a firm for a given product or brand can be grasped from the target market chosen, the way it is positioned and how the marketing mix is organised. The target market shows to whom the unit intends to sell the products; positioning and marketing mix show how and using what uniqueness or distinction, the unit intends to sell. These three together constitutes the marketing strategy platform of the given product.

1. Selecting the target market: it shows to whom the unit intends to sell the product. Target market selection is a part of marketing strategy development. When the selection of the target market is over, an important part of the marketing strategy of the product is determined, defined and expressed.

   e.g. Reliance Textile, part of Reliance group entered the Indian Textile market in 1967, it found that this market considered of many distinct segments, spread over entire rural and urban India. It was a Rs.5,000 crore market, with cotton textile taking more than 70 per cent share and the rest taken up by silk and synthetics. Reliance was coming out with a premium product-high quality synthetic fabrics, sarees, suitings, shirts and dress materials. Reliance had to select its target market for this product. It spotted the well-to-do and fashion loving upper middle class of urban India as its target market. The decision came through a combined process of analytical exercise and executive judgement.

2. Positioning : The firm has already selected the target market and decided its basic offer. Now what is the conjunction between these two entities? How do they get connected ? What is the interface? What is the locus the firm seeks among the customers in the chosen target market with its offering? How would the firm want the consumer to view and receive the offers? What position it seeks and what image it proposes to build for its offers(most prestigious offer and most competitive offers etc.)?

3. Assembling the marketing mix: Marketing mix is the sole vehicle for creating and delivery customer value.

Marketing Mix

The four elements or 4P's of marketing constitute the **marketing mix.**

i) Product- (design, features, brand, model, style, quality, package service etc.)

ii) Place- (channels of distribution, types of intermediaries, location of outlets, physical distribution, transportation, warehousing, inventory etc.)

iii) Price- (policies, margins, discounts, rebates, payment terms credit terms, resale, price maintenance etc,)

iv) Promotion- (personal selling, advertising, publicity and public relation etc.)

The 4P's are best possible combination which involved in the process of choice of the appropriate marketing activities and allocation of the appropriate marketing efforts and resources to each of them.

Marketing mix

Four P's of marketing mix:

Marketing management

According to Philip Kotler, Marketing management is the analysis, planning, implementation and control of programmes designed to bring about desired exchanges with largest audiences for the purposes of personal and mutual gains. It relies heavily on the adaptation and coordination of product, price, promotion and place for achieving effective response.

Marketing management is the art and science of choosing target markets and getting, keeping and growing customers through creating, delivering and communicating superior customer value.

Marketing management is the conscious effort to achieve desired exchange outcomes with target markets.

Marketing management refers to the planning, organizing, directing and controlling the activities of a person working in the marketing division of a business enterprise with the aim of achieving the organizational objectives of : (i) achieving higher productivity in its marketing operation by making optimum use of the available resources, (ii) enhancing the productivity of enterprise with the consumer orienting marketing.

Marketing management	V/s	Strategic management
Action oriented		Analysis oriented
Existing opportunities		New opportunities
Non-product variables		Product markets
Stable environment		Dynamic environment
Reactive behavior		Pro-active behavior
Day-to-day management		Long-range management
Marketing department		Cross functional organization
Low diversification		High diversification
Profit objectives		Customer satisfaction

Objectives

1. Increase the state volume
2. Increase in net profits
3. Growth of enterprise

Responsibilities of the marketing manager

1. Market analysis
2. Determining market goals
3. Organising the marketing activities
4. Assembling
5. Product management
6. Controlling marketing activities

Managerial functions in marketing

1. Determining the marketing objectives
2. Planning
3. Organization and coordination
4. Staffing
5. Operation and direction
6. Analysis and evaluating

Demand

It is a schedule that shows the amounts of product or service the consumers are willing and able to purchase at each price in set of possible prices during some specified time in a specified market.

Consumers like to possess a particular commodity but without ability to pay, in which case it is not a demand. Apart from these, two more requisites are essential viz. time and market, for demand is likely to vary over time and also among the markets. The conditions hence imposed are specific time and a specific market to measure the demand. One condition that governs here is that all other factors, other than price influencing the demand remain the same.

According to Bowden, demand means propensity of the consumers to buy different quantities of a particular good at different unit prices. It indicates how much the consumers would be buying when the price per unit of a commodity is changing.

Individual demand schedule

The various quantities of a commodity that a consumer would be willing to purchase at all possible prices in a given market at a given point in time other things being equal is called individual demand. Individual demand schedules merely a list of prices together with the quantities that will be purchased by a consumer. It is pairing of price and quantity relationship.

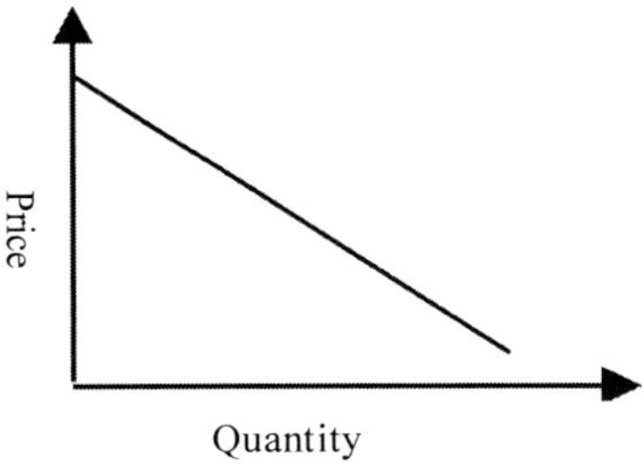

Fig. Demand curve

Market demand

Market demand is the sum of the demand of all the consumers in a market for a given commodity at a specific point of time. Or the market demand for a given commodity is the horizontal summation of the demand of the individual consumers. Or the quantity demanded in the market at each price is the sum of the individual demands of all consumers at that price.

For example, three consumers viz. A, B and C.

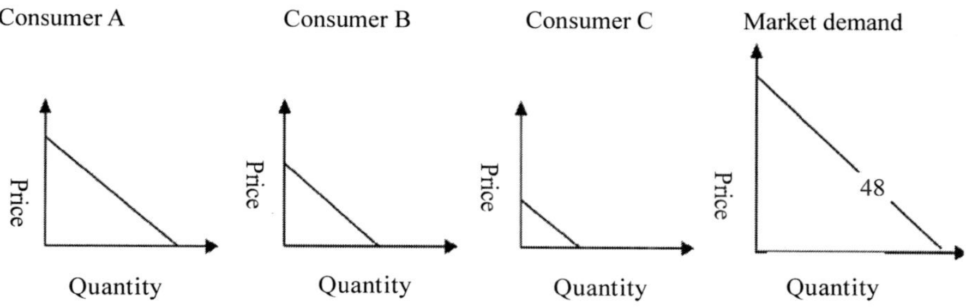

Fig. Market demand curve

Autonomous demand and derive demand

Autonomous demand

The goods whose demand is not linked with the demand of other goods are supposed to have autonomous demand. Consumer goods are the example here.

Derive demand

The demand for certain goods is related with the demand for other goods which is called derived demand. The demand for fertilizers, pesticides etc is a derived demand for it is linked with the demand for agricultural product. Thus the goods which are demanded for their own sake have autonomous demand while the goods that are required to produce other goods have derived demand.

Kind of demand

1. **Price demand:** It refers to various quantities of a good or service that a consumer would be willing to purchase at all possible prices in a given market at a given point in time, ceteris paribus.
2. **Income demand:** It refer to various quantities of good or service that a consumer would be willing to purchase at different levels of income ceteris paribus.
3. **Cross demand:** It refers to various quantities of a good or service that a consumer would be willing to purchase not due to changes in the price of the commodity under consideration but due to changes in the price of related commodities.

Law of demand

The law of demand explains the functional relationship between the quantity demanded of a commodity and its unit price i.e., a rise in the price of a commodity or service is followed by a reduction in the quantity demanded and a fail in the price is followed by an extension in demand with other condition remaining the same. Increased prices tend to contract demand andfalling prices extend it. Demand various inversely with the price other things being equal. The tendency of the consumer is to buy more quantities of a commodity at a lower price than what he buys at a higher price.

In short, quantity demanded is inversely related with the price i.e.,

$$D \propto \frac{1}{P}$$

Where,

D = Demand

P = Price

$\propto$ = inversely

Exception to the law of demand

1. **Giffen goods (inferior goods):** This phenomenon which is explained below was given by Sir Robert Giffen. It was name after him as Giffen paradox. This phenomenon says that raise in price is followed by an extension of demand, while a fall in price is followed by a reduction in demand for the food. In the case of poor who spends major portion of their income on an inferior commodity like Bajra, are left with lesser amount to spend on other items. Suppose in a situation where the price of Bajra with the prices of other commodities and money income remaining the same, the rise in price of Bajra makes him worse-off than before. Further, he cannot buy the same quantity as he was buying earlier with allotted money. He cannot afford to spend the same amount which he was earlier allotting, lest his family will be starved. Now he has to cut down the expenditure on other items not only to maintain but also to increase the quantity of Bajra he bought per unit of time. On the other hand if the price of Bajra falls the increased real income enables him to spend on superior foodgrains. This leads to a contraction of demand for Bajra, even though there is a fall in its price.
2. **Prestigious goods:** When the possession of a good brings in social distinction, consumer would go for the same event if its price is higher: Example to be cited here is the diamonds that the rich people purchase, as the possession of the same is prestigious to them.
3. **High priced commodities:** When the consumers view that those products which are superior are sold at higher prices, they do not mind to buy more of the same at higher prices.

4. **Fear of shortage:** If the existing price is higher and it is expected to increase further, consumers would buy more of it even at higher price, fearing for the shortage.

Movement along the demand curve

It refers to change in the quantity demanded due to change in price. it can be either extension or contraction of demand.

Extension of demand means buying more quantity of commodity at a lower price, while contraction of demand indicates buying less at a higher price.

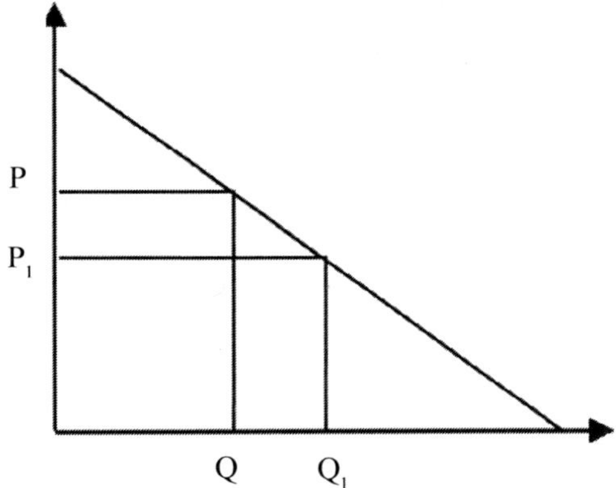

Fig. Movement along the demand curve

Shifting in the demand curve

It refers to change in demand not due to change in price but due to change in the values of other variables influencing demand. It can increase or decrease in demand. As against extension and contraction of demand, increase and decrease in demand result in the shifting of the demand curve. Increase in demand means more demand at the same price or same demand at higher price. On the other hand, decrease in demand means less demand at the same price or same demand at lower price. When there is increase in demand, the demand curve shifts upward to the right side of the initial demand curve DD. D_1D_1 is the new demand curve representing increase in demand. At 'OP' price, the quantity demanded is OQ. Increase in demand indicates the purchase of same quantity of a commodity (OQ) at a higher price of OP_1 or purchase of OQ_1 quantity at the same price of OP. the decrease in demand is indicated by the shift of the demand curve towards left downwards to the initial demand curve. D_2D_2 is the demand curve representing decrease in demand. Decrease in demand indicates purchase of the same quantity ofa commodity (OQ) at a lower price of OP_2 or purchase of OQ_2 quantity at the same price of OP. Increase and decrease denote change in demand.

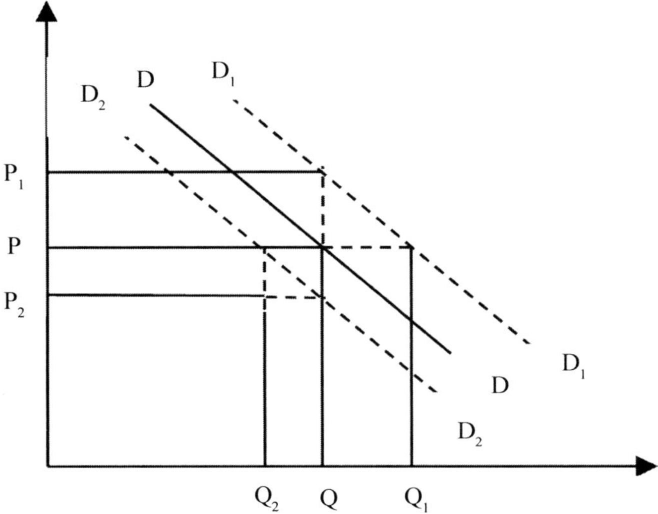

Fig. Shifts in the demand curve

Factors affecting demand (shift factor)

The change in demand for a produce is due to the factor other the price factor of the same product. These are tasted and preferences, income of the consumer, prices of related commodities etc.

1. **Tastes and preferences:** These are influenced by fashions, population changes, advertisement, customs, habits, popularity, season etc. changing fashions in the men/women wear, ornaments etc will attract the people to buy them, so that they are fit to the changing fashion world. Changes in population also bring variations in their tasted and preferences. New generation will have something different attitudes in all walks of life, consequently new tastes emerge. Advertisements has become an important instrument of sales promotion/selling. Strategies of the manufactures. With the growing access to several mass media sources, people are beholden by their impact, consequently tastes and preferences are ever changing. Customs of the society enforce the people to change their life styles, as a result of which new preferences add to their long list of wants. People are habituated to certain aspects of life and develop new preferences. Popularity of a

product attracts the new consumer towards it. This implies that we have new preferences for products. Popularity of the product is one, the influence of which cannot be a avoided by the consumer. Season-bound requirements compel the people to have a new set of preferences each season. Thus, the changes in tastes and preferences make the consumer more satisfied than before.

2. **Income:** Changes in the income of the consumer leads to s shift in the demand curve. The proportion of income spent on food and other necessaries decreases with increase in the income level of the consumers.
3. **Price of other related goods:** The change in price of one good influences the consumption of other good, depending upon the relationship between the goods. If two commodities are supposed to be substitutes such as meat and chicken, coffee and tea etc the rise in price of one goods which need to be used together viz bread and jam, bread and butter etc which are called complements. Increase in price of one good brings down the demand for the other good. A rise in price of bread reduces the demand for butter.
4. **Habits:** Large numbers of smokers in a region influence the demand for tobacco and its products.
5. **Region:** In high altitude regions demand for wine, woolen clothing, meat etc would be persistent.
6. **Season:** Demand for eggs decreases in summer season and increases in winter season.

Elasticity of demand

The law of demand says that demand varies inversely with the price, other things being equal. From the law of demand, we know the direction in which quantity and price are moving. What is not known is the extent by which is precisely needed by responsive to changes in price. This is the information, which is precisely needed by businessmen and policy makers. Alfred Marshall developed the concept of elasticity of demand which measures the responsiveness of quantity demanded to changes in price. Elasticity of demand indicates the degree of relation between quantities demanded changes becauseofchange in price. To be more precise, elasticity of demanded is defined as "the relative change in the quantity demanded to the relative change in the price".

Types of elasticities of demand

There are three types of elasticities of demand viz.

i) Price elasticity of demand

This shows the responsiveness of quantity demanded of a commodity, when price of that commodity change, with other factors being constant.

Price elasticity of demand (E_d) $= \dfrac{\text{\% change in quantity demanded}}{\text{\% change in price}}$ (OR)

$$= \frac{\text{Proportionate change in quantity demanded}}{\text{Proportionate change in price}} \quad \text{(OR)}$$

$$= \frac{\dfrac{\text{Change in quantity}}{\text{Initial quantity}} \times 100}{\dfrac{\text{change in price}}{\text{Initial price}} \times 100}$$

ii) Income elasticity of demand

It measures the responsiveness of demand due to changes in the income of the consumers in terms of percentage, when other factors influencing demand viz. price of the commodity, price of substitutes, tastes, preferences etc are kept at constant level.

Income elasticity demand $= \dfrac{\text{\% changes in quantidy demanded}}{\text{\% change in income}}$ (OR)

$$= \frac{\text{Proportionate change in quantity demanded}}{\text{Proportionate change in income}} \quad \text{(OR)}$$

$$= \frac{\dfrac{\text{Change in quantity}}{\text{Initial quantity}} \times 100}{\dfrac{\text{change in income}}{\text{Initial income}} \times 100}$$

iii) Cross elasticity of demand

Demand for one good (X) is also influenced by the price of other related good (Y). These may be substitutes or complements. It is the ratio of percentage change in quantity demanded of commodity (X) and percentage change in price of related commodity (Y).

Cross elasticity of demand (E_{xy})

$$= \frac{\text{\% changes in quantidy demanded of commodity (X)}}{\text{\% change in price of related commodity (Y)}} \quad \text{(OR)}$$

$$= \frac{\text{Proportionate change in quantity demanded of commodity (X)}}{\text{Proportionate change in price of related commodity (Y)}} \quad \text{(OR)}$$

$$= \frac{\dfrac{\text{Change in quantity}}{\text{Initial quantity}} \times 100}{\dfrac{\text{change in price}}{\text{Initial price}} \times 100}$$

Degree of elasticity of demand

Based on the magnitudes of elasticity of demand, it can be categorized into five degree viz. perfectly elastic demand, perfectly inelastic demand, relative elastic demand, relative inelastic demand and unitary elastic demand.

Perfectly elastic demand

A slightest change in price of a commodity leads to an infinite change in quantity demanded. The demand in such a situation is said to be perfectly elastic. The demand is and the elasticity of demand is infinite. Here the demand curve will be horizontal line parallel to X-axis. It is a theoretical concept.

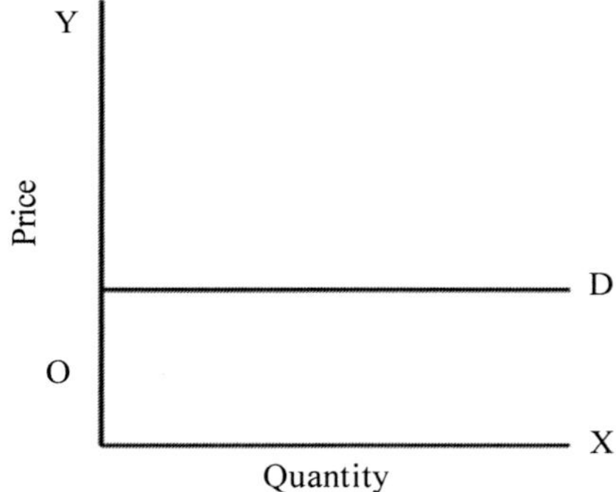

Perfectly inelastic demand

It is situation in which change in price of a commodity leaves the demand unaffected. The price of the commodity may increase or decrease, but the quantity demanded remains the same. The demand here is insensitive. Elasticity of demand is zero. The demand curve is vertical to X-axis. It is also a theoretical concept.

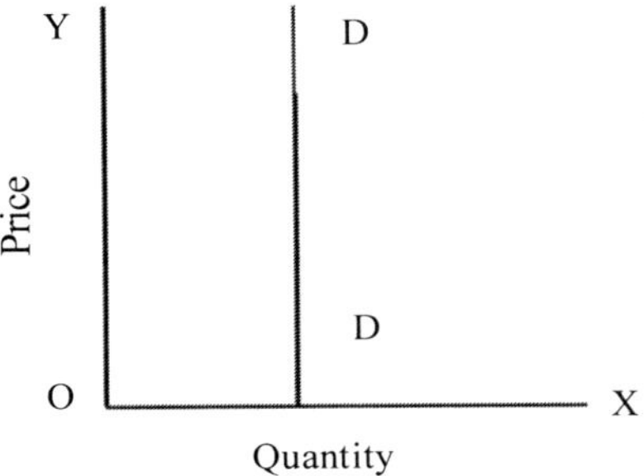

Relative elastic demand

It means that lesser proportionate change in the price of a commodity is followed by a larger proportionate change in the quantity demanded. A small proportionate fall in the price is accompanied by a larger proportionate increase in demand and vice versa. Elasticity of demand is greater than unity.

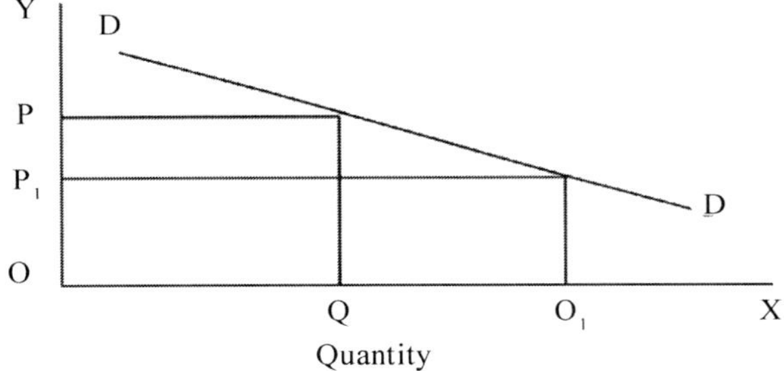

It means that large proportionate change in the price of a commodity is followed by a smaller proportionate change in the quantity demanded. This is to say that a large proportionate fall in the price is followed by a smaller proportionate increase in the quantity demanded and vice versa. Elasticity of demand is less than unity.

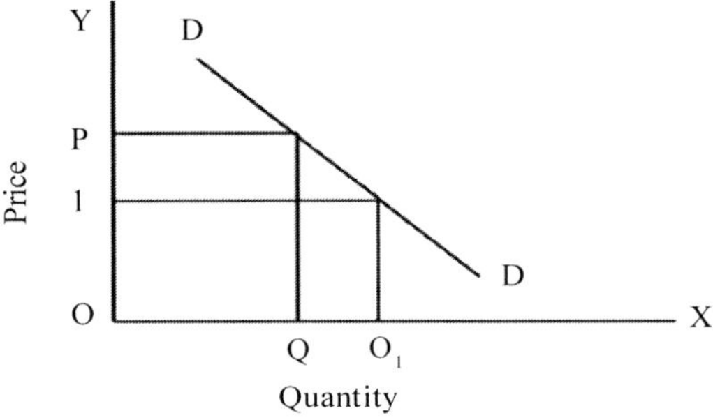

Unitary elasticity demand

When a given proportionate change in price results in the same proportionate change in the quantity demanded of a commodity, the demand is said to be unitary elastic. A given proportionate fall in price is followed by the same proportionate increase in demand and vice versa. Elasticity of demand is one.

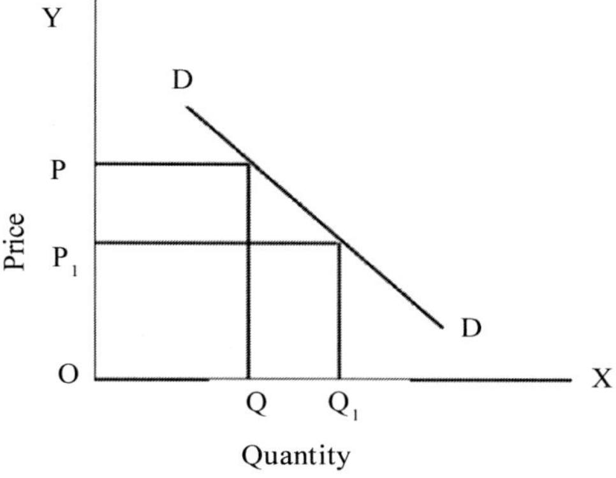

Practical importance of elasticity of demand

1. **Determination of wages**: The elasticity of demand for labour plays an important role in the determination of wages. If the demand for labour is elastic, any pressure put up labour in the form of strikes to get higher wages would be unsuccessful. On the other hand, if the demand for labour happens to be inelastic, even a threat of strikes would help the workers to get the approval of their employers in raising their wages.
2. **The elasticity of promotional activity:** The producers are well convinced that the advertisement makes the demand for a product less elastic. Hence, they would not mind spending substantial amount of money on advertising. The price increase therefore will not reduce the sales.
3. **Determination of Monopoly price:** The monopoly considers the nature of demand for his product before fixing the prices. If the demand is elastic, a lower price would help him to realize more profits. On the other hand, if the demand is inelastic, he is in a position to fix a higher price. The monopolist while practicing price discrimination also takes into account the elasticity of demand. He fixes a lower price for the product in the market in which the demand for the product is elastic and he charges a higher price for the same product in the market in which the demand is less elastic or inelastic.
4. **Undertaking the public utilities:** The government itself runs some enterprises or industries in the interest of people, otherwise they are subjected to exploitation by the private people. In the case of electricity, the demand is

inelastic and it is very essential item in the lives of humans. In the interest of the public, electricity boards are run by the government to supply power at reasonable rates.

5. **Taxation:** The nature of demand for a good helps the government, looking into the possibilities of raising the revenue. If the demand is less elastic for a good, by levying more indirect taxes on that good, the government would get larger revenue.
6. **International trade:** The country gains in the international trade by exporting those goods, for which the demand in the export market is less elastic and by importing those goods for which the demand is elastic. The less elastic nature of demand in the export market helps the nation to charge a higher price and pay less price for those goods which are imported and
7. **Paradox of poverty in plenty:** A bumper crop instead of bringing prosperity to the farmers, ruin their economic position. It is common phenomenon in agriculture. The inelastic demand for the products in the years of bumper harvest brigs down the prices, consequent to which the farmers fail to realize prices of normal years.

Methods of measuring price elasticity

1. Total outlay or expenditure method

In this method we compare total expenditure of the consumer before and after change in price. The elasticity of demand is unity when the total expenditure remains unaltered even though, there is price change. The demand is said to be elastic when the total expenditure increases with fall in price and decreases with rise in price. Inelastic demand is observed when the total amount spent on the commodity by the consumer increases with increase in price and decrease with fall in price.

Between 1 and 2, the elasticity is greater than unity (relative elastic) because the total expenditure increases with fall in price and decreases with rise in price. between 3 and 4, the elasticity is unitary as the total outlay remains the same, though there is variation in the price. between 5 and 6, the fall in price is followed by a fall in the total outlay and rise in the price is accompanied by an increase in total expenditure. Therefore, the elasticity is less than unity (relatively inelastic).

2. Point elasticity of demand

It is a geometrical method for measuring elasticity at a point of the demand curve. Point method is used when price and quantity changes are extremely small. This method is applicable only when we possess information on the minutest changes in price and quantity.

A straight-line demand curve joining the two axes in considered showing the measurement of point elasticity. Elasticity at any point on the demand curve is the ratio of lower part of the straight line to the upper part. It is important to note that point elasticity of demand on straight line is different at every point. At any point to the right of midpoint (C) the point elasticity is less than unity, at any point to the left of midpoint the point elasticity is more than unity and at mid point (C) the elasticity is unity. At the point where the linear demand curve intersects Y-axis (A) the point elasticity is infinite, while at the point where the demand curve intersects X-axis (E), the point elasticity is zero.

3. Arc elasticity of demand

The price elasticity of demand measured between two distinct points on a demand curve is called arc elasticity of demand. This method uses the mid points between the old and new data in the case of price and quantity. It studies a portion of the demand curve between two points. An arc is a segment or a portion of a curved line. Arc elasticity is employed in order to compute price elasticity coefficient from the discrete data. The following formula is used to compute are elasticity.

$$E_d = \frac{\dfrac{Q_1 - Q_2}{\dfrac{(Q_1+Q_2)}{2}}}{\dfrac{P_1 - P_2}{\dfrac{(P_1+P_2)}{2}}} = \frac{\dfrac{\Delta Q}{(Q_1+Q_2)}}{\dfrac{\Delta P}{(P_1+P_2)}}$$

Where,

Q_1 = first quantity observed

Q_2 = second quantity

P_1 = first observed price

P_2 = second price

Factor determining elasticity of demand

1. **Type of goods:** The demand is inelasticity or less elastic for necessaries. It is obvious because the price changes do not influence the consumption of these goods. On the other hand, I respect of comforts and luxuries, the demand is elastic or relatively elastic.

2. **Goods having several uses:** The demand for those goods which can be put to several uses is elastic. For example, electricity when it is cheap, it can be put to uses like cooking, apart from its regular use in production of industrial goods, transport, lighting etc. In case it is dearer it use is limited and demand automatically declines.

3. **Existence of substitutes:** The demand for those commodities which have good substitutes is elastic. The price variations of such goods would have a bearing on the quantities demanded. If the price of groundnut oil increases, with the prices of other edible oils, say sunflower remaining constant, the consumers tend to shift to sunflower oil. Conversely a decrease in price of groundnut oil makes consumers to switch over back groundnut oil.
4. **Possibilities of postponement:** For such goods the purchase of which can be postponed by the consumers, the demand is elastic. The consumers prefer to buy the same when their price is cheap. Televisions, care, ornaments etc are the relevant examples here.
5. **Range of price:** When the prices ranges of goods are very high or very low the demand is inelastic. The demand for goods like salt, matches etc with low price range, and the demand is inelastic. Analogously for goods like diamonds, luxurious cars etc the demand is inelastic as consumer cannot afford to the changes in the price of such a high range.

Supply

Supply means the producer offered the quantity of a good or service for sale at different unit prices in a given market at a point of time. It is the supplier willingness to offer the goods for sale at different unit prices. It also can be defined as a schedule that shows the amounts of a product or service, sellers are willing to sell at each unit rice in a set of possible price during some specified period of time in a specified market.

Meyers defined supply as "a schedule of the amount of good that would be offered for sale at all possible prices at any one instance of time in which the condition of supply remains the same".

Prof. Mc. Connel defined supply as "a schedule which shows the various amounts of a product which a producer is willing to and able to produce and make available for sale in the market at each specific price in a set of possible prices during some given period".

Individual supply schedule

Supply schedule shows the list of relationships between quantities-price of a commodity in a market at a specific point of time by an individual seller. In other words, it reveals the mind of sellers in offering various quantities of a given commodity against corresponding prices.

The relation between supply and price of the commodity are directly proportional to each other. Hence, as the price per unit of the commodity rises, the quantity supplied is also increasing. As price increases, sellers are committed to increase their sales. When a supply schedule is plotted on a graph it becomes a supply curve. The supply curve will have a positive slope i.e., it slopes upwards from left to right.

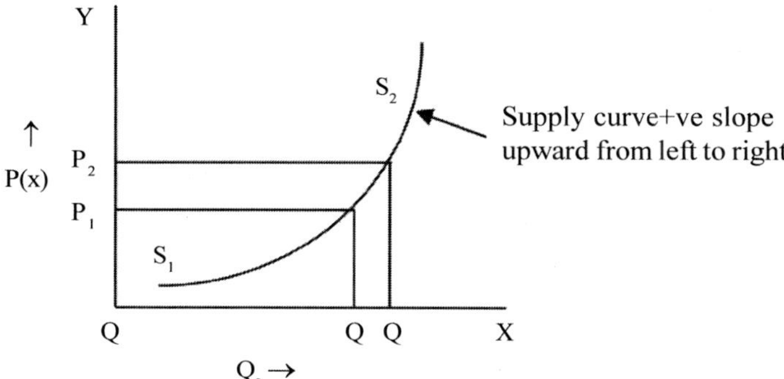

Fig. Supply curve

Market supply

It is the sum of the quantity of commodity that is brought into a market for sale by the sellers in a given market at a specific point of time. Assume that there are three sellers in a given market viz., A, B and C with individual supply schedules as shown in table

Market supply schedule

Price(in Rs./Q	Individual seller's supply/week (Q)			Market supply (Q) = (A+B+C)
	A	B	C	
300	30	25	0	65
325	40	50	0	90
350	50	65	50	165
375	60	80	70	210
400	70	95	90	255
425	80	110	110	300

The price quantity relationship of the three sellers reveals that at Rs. 300 per quintal, seller 'A' is prepared to sell 30 Q, while seller 'B' 35 Q and seller 'C' is not prepared to sell at all at this particular price. The seller 'C' is not prepared to sell the commodity at any price less than Rs. 350/Q. market supply is the sum total of output that is sold by the three sellers as presented in the last column of the table. Thus the market supply is 65, 90, 165 Q and so on. It is the lateral or horizontal summation of the supply of individual sellers at each unit price. The market supply curve is drawn based on the first and last columns of the table. The graphical presentation of market supply is

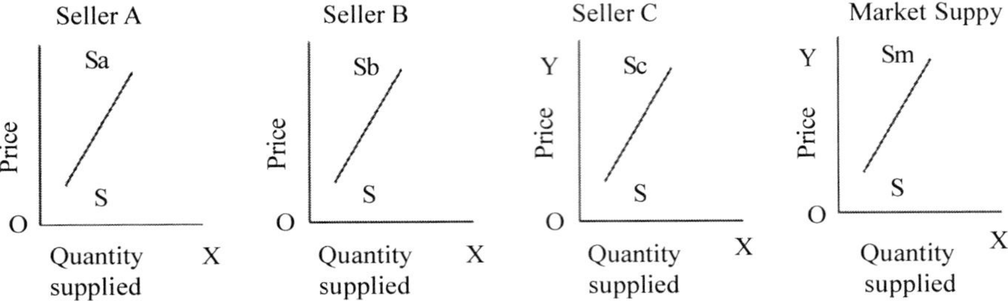

Law of supply

The law of supply indicates the functional relationship between the quantity supplied of a commodity and its unit price. The law signifies the positive relationship i.e., as the price of a commodity rises its supply extends and as the price falls its supply contracts, with other things remaining the same. Producers normally tend to increase the supplies in the wake of rising prices and reduce the same when the prices are on the lower side. Supply varies directly with the price, ceteris paribus.

Extension and Contraction of Supply (Change in Quantity Supplied)

Extension and contraction of supply refer to the movement of product supply on the same supply curve. Extension of supply means offering more quantity for sale at a higher price, while contraction means offering less quantity at a lower price. As is seen from table that the quantity of commodity supplied by 'A' at Rs. 300 is 30 Q and it is 40 Q when the price rose to Rs. 325. Here the quantity supplied has increased from 30 to 40 Q.

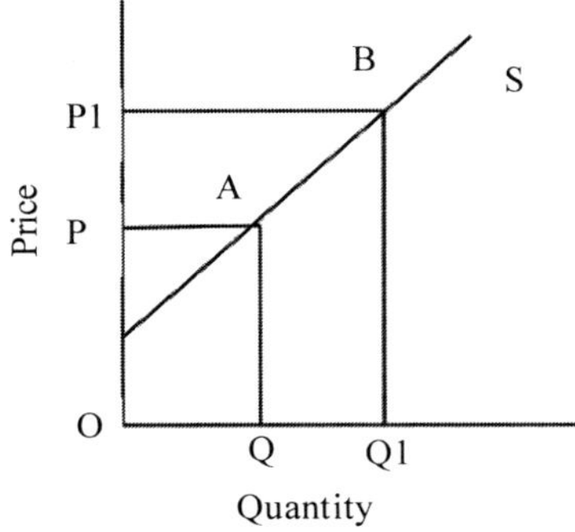

Fig. Extension and contraction of supply

It is the case of extension of supply, conversely if the price falls from Rs. 325 to Rs. 300, the quantity supplied to falls from 40 to 30 Q. It is the contraction of supply.

Geographically, it shows that the upward movement from A to B is extension of supply and downward movement from B to A is contraction of supply.

Increase and Decrease in Supply (Shift in Supply)

Increase in supply implies more supply at the same price and decrease in supply means less supply at the same price. The change in supply (increase and decrease in supply) results in a shift of the supply curve. An increase in supply results in the shift of the supply curve towards right side of the initial supply curve SS. The new supply curve is S_1S_1. On the other hand, a decrease in supply causes a shift of the supply curve towards the left side of the initial supply curve. The new supply curve thus formed is S_2S_2. Originally OQ quantity is supplied at OP price. But due to changes in supply conditions at the same price OP, OQ_1 quantity f commodity is supplied indicating increase in supply. On the other hand, again influenced by changing supply conditions at the same price, OQ_2 is supplied. This is decease in supply.

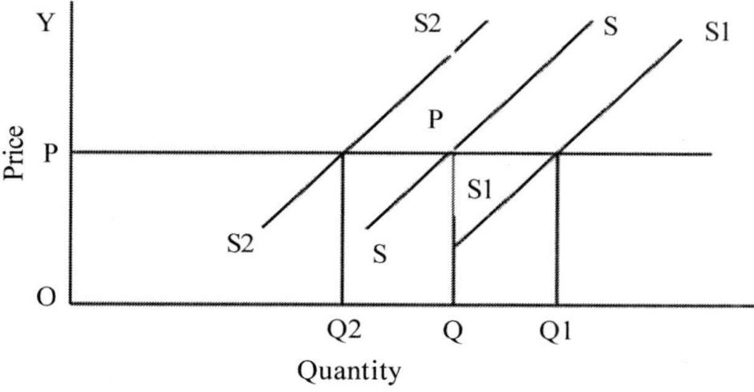

Fig. Increase and decrease in supply

Factor Causing Changes in Supply (Shift Factors)

The factors that are responsible for changes in supply are discussed below:

1. **Changes in technology**: Technological innovations viz., new varieties of crops and their consequent increased yields per unit area, help to increase the supply of the commodity.
2. **Reduction in resource prices**: When the prices of input factors become cheaper than before, it encourages producers to use more of them in producing more output. Supply curve shifts towards the right side.

3. **Reduction in the relative prices of other products**: A reduction in relative prices of other related products compel the producers to increase the production of that particular commodity whose prices are relatively higher.
4. **Market infrastructure**: When good communication and transport network increase the supply of the commodity also increases.
5. **Number of producers**: Changes that are found regarding number of producers producing a given commodity influence the supplies. More the number o producers, greater the supply and vice versa.
6. **Producer's expectations about future prices**: Price expectations influence the sales strategies of the producers positively.

ELASTICITY OF SUPPLY

Elasticity of supply of a commodity is the responsiveness or sensitiveness of supply to the changes in price. Supply is said to be elastic, if a small change in price causes considerable change in the quantity supplied. The supply is inelastic when a given change in price leads to little or less change or no change in the quantity supplied. In short, elasticity measures the adjustability of supply of a commodity to price.

Elasticity of supply (price elasticity of supply) is expressed as the ratio of percentage change in the quantity of good supplied and percentage change in price of the good ceteris paribus.

$$\text{elasticity of supply} = \frac{\text{Percentage change in quantity of good supplied}}{\text{Percentage change in price of good supplied}}$$

Algebraically elasticity of supply is expressed as:

$$\frac{\Delta Q}{Q}\times 100$$

$$\frac{\Delta P}{P}\times 100 = \frac{\Delta Q}{Q}.\frac{P}{\Delta P} = \frac{\Delta Q}{\Delta P}.\frac{P}{Q}$$

Degrees of elasticity of supply

There are five degrees of elasticity of supply. They are as follows

PERFECTLY ELASTIC SUPPLY

When the supply of commodity increases to infinite quantity or unlimited quantity even though there is invisible rise or minute rise in the price the elasticity of supply is said to be infinity i.e., $E_s = \alpha$

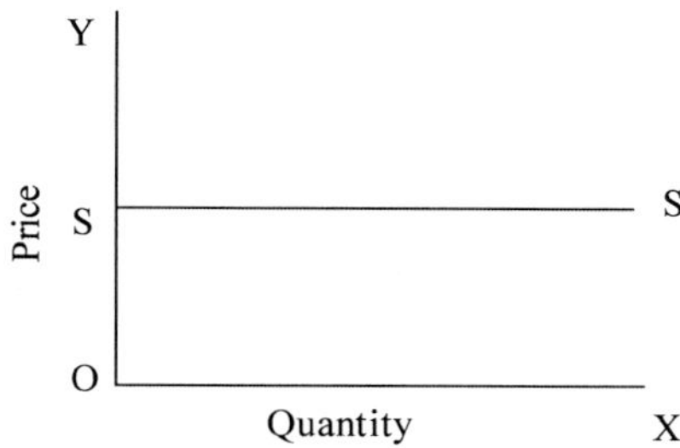

Fig. Perfectly elastic

PERFECTLY INELASTIC SUPPLY

It means that the quantity supplied is not responsive to change in prices. Elasticity of supply in this case is zero ($E_s = 0$).

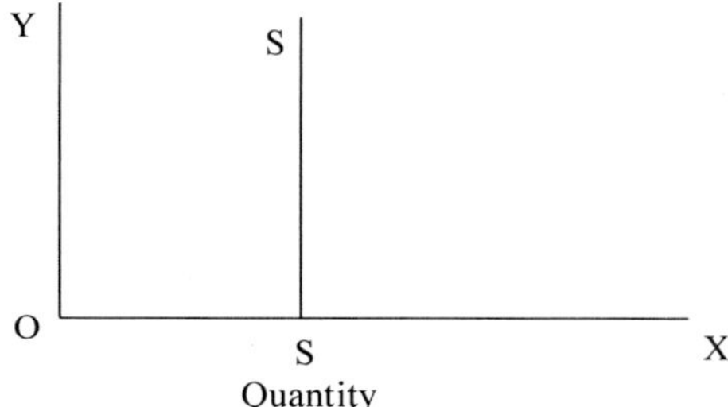

Fig. Perfectly inelastic

RELATIVELY ELASTIC SUPPLY

Supply is referred as relatively elastic, when the percentage change in quantity supplied is more than the corresponding percentage change in price. It is also called elastic supply. Elasticity of supply is more than one (E_sÃ 1).

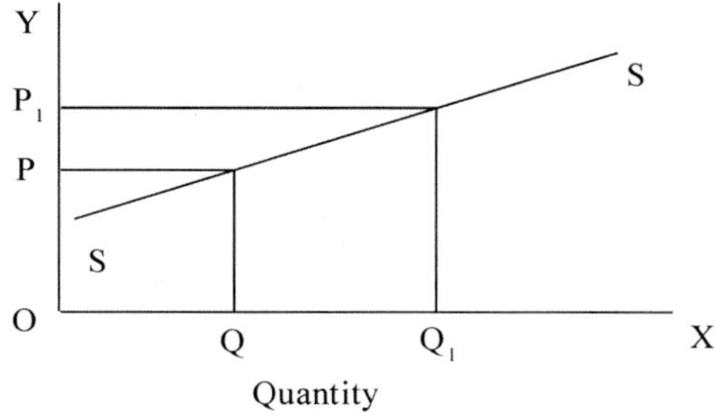

Fig. Perfectly inelastic

Relative Inelastic Supply

Supply is said to be relatively inelastic, when the percentage change in quantity supplied is less than the corresponding percentage change in price. In this case the elasticity of supply is less than one (E_s<Â1).

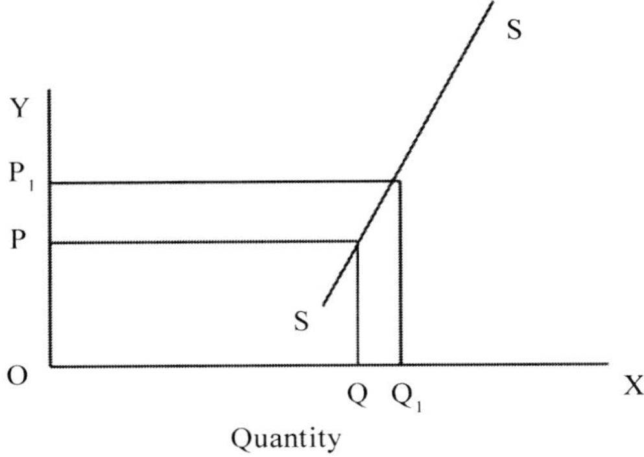

Fig. Perfectly inelastic

Unitary Elastic Supply

When percentage change in quantity supplied equals the percentage change in price, it is called unitary elastic supply. Here the elasticity of supply is equal to one ($E_s = 1$).

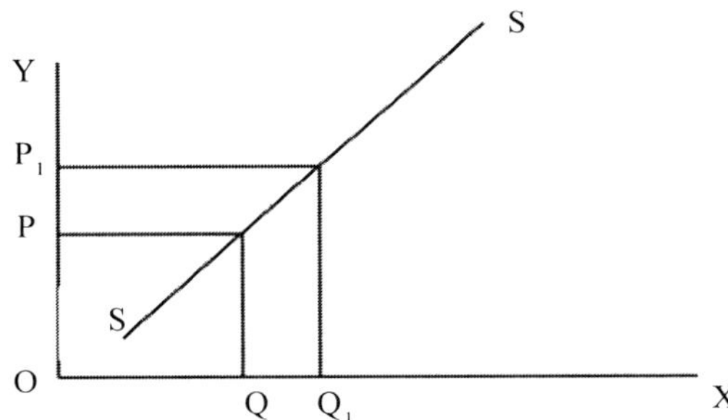

Fig. Perfectly inelastic

Factor Influencing Elasticity of Supply

1. ***Availability of Inputs of Production***: If the needed in puts are available as per the requirement, the supply is elastic. If any one of the factors is not available which is absolute necessary, supply would be inelastic.

2. ***Length of Time Period***: it is the period of time required to adjust the supplies to the changes in prices. The biological characteristics of the product dictate the changes of responsiveness.
3. ***Diversification of production activity***: When the producer is engaged in production of a number of products and facilities exist for shifting of production from one product to the other, in such a case for each product the supply is elastic.
4. ***Availability of Alternative Markets***: Suppose there exists several markets for the producer to sell the goods, a fall in price in one market would prompt him to shift his gods to other markets and a rise in price in one market induces him t shift his goods to that market. In such a case the supply is elastic.
5. ***Flexibility in Starting and Winding up the Business***: If a particular production activity is quickly taken up and quickly wound up, the supply of the goods is elastic.

PRICE DETERMINATION

Having studied the demand and supply, we now that market demand curve is a horizontal summation of individual demand curves, and similarly horizontal summation of individual supply curves become market supply curve.

Price determination can be examined arithmetically, graphically and algebraically.

Arithmetic Approach

The information in table reveals that at Rs. 12, the quantities demanded and supplied are both equal i.e., 80 Q. At this price, what the buyers are willing to purchase and what the sellers are willing to offer are the same. Therefore, Rs. 12 per unit is the equilibrium price and quantity amounting to 80 Q is the equilibrium output.

Table: Demand, Supply and the Equilibrium Price and Output.

Price/ unit (Rs)	Quantity demanded (Q)	Quantity supplied (Q)
14	60	120
13	70	100
12	80	80
11	90	60
10	100	40

Graphic Approach

The intersection of market demand curve (DD) and the market supply curve (SS) indicates the equality of quantity demanded by the consumers and that supplied by the producers in fig. This equality of quantity demanded and quantity supplied is called equilibrium quantity (OQ) and the price that occurs at this balancing point is called equilibrium price (OP). when such a condition prevails in a market, the market is said to be in equilibrium, because there are either shortages nor surpluses of commodities in fig.

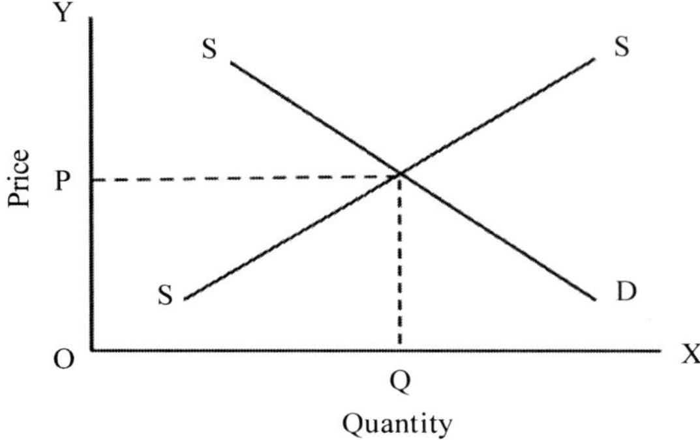

Market Disequilibrium

As observed in fig, at OP_1 price suppliers are prepared to sell OQ_2 quantity, while the consumers prefer to buy OQ_1 quantity only. Hence OP_1 price is not in equilibrium as quantity to the extent of $(OQ_2\text{-}OQ_1)$ is available as surplus. Since there is a surplus, the market is in disequilibrium. Suppose if the seller wants to sell the surplus they have to oblige the pressure of consumers to reduce the price. When the same takes place i.e., when price slides down to O the consumers are prepared to purchase what all the suppliers offer for sale.

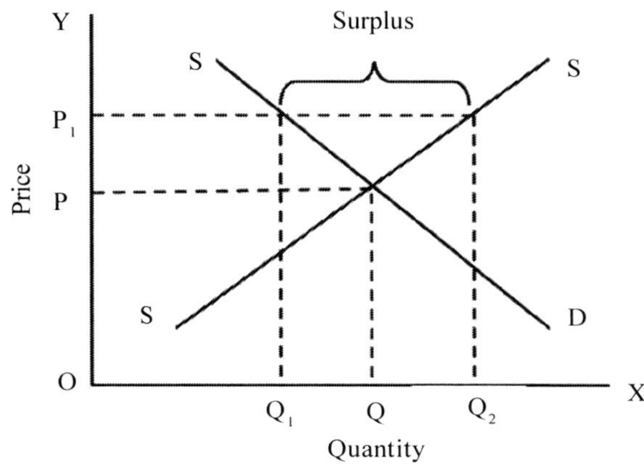

A Case of Shortage

At price PO_1, which is lower than the equilibrium price, the suppliers are prepared to sell smaller quantities depicted as OQ_1 in the fig. The quantity demanded by the consumers on the other hand is OQ_2 quantity, which is higher than the quantity the sellers prefer to sell. The gap between OQ_1 and OQ_2 indicates the shortage of the commodity in the market. Since the quantity supplied is less than the demand in the market, the consumers are willing to get the quantity they required consequent to which the price rises up to OP, which is an equilibrium price. At the price OP, the quantity supplied by the suppliers equals the quantity that the consumers demand i.e., OQ.

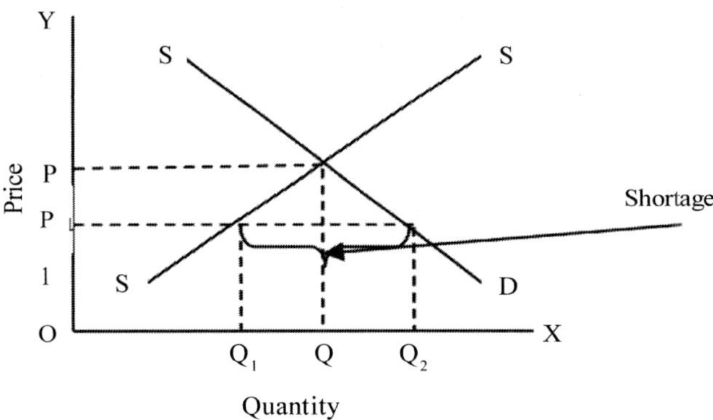

Estimation of Elasticity of demand of Horticultural produce.

Price Elasticity

It is the relationship of the changes in quantity to changes in price. It is the change in commodity price. It is the proportional change in quantity demanded by the proportional changes in price.

Elasticity of demand

$$Ed \frac{\Delta q}{\Delta p} \times \frac{p}{q}$$

$$= \frac{\text{\% change in quantity demended}}{\text{\% change in price}}$$

Ed > = Elastic

Ed = 1 = Unities

Ed < 1 = Inelastic

Example

Supposed the price of Khasi mandarin in a market is Rs.80 per Kg and its demand in same market is 1000 kg as a result of price rising in 20% its demands has fallen to 900 Kg. calculate the elasticity of demand (Ed).

Here,

$\Rightarrow P_1 = \text{Rs.}80$

= Rs. 80 of 20% rise in price

$= 80 \times \frac{20}{100}$

$= 16$

$\Delta q = q_2 \sim q_1 = 1000 \sim 900$

$= 100$

$\Delta P = 16$

$\therefore ED = \frac{\Delta q}{\Delta p} \times \frac{P}{Q}$

$= \frac{100}{16} \times \frac{80}{1000}$

$= 0.5$

$\because ED < 1$

$\therefore$ it is inelastic

5

Stakeholders Analysis

Stakeholder is an individual or group that has an interest in any decision or activity of an organization. A stakeholder is a member of groups without whose support the organizationwould cease to exist. Stakeholders are important to every project. Stakeholders are very important because they can have a positive or negative influence on the project with their decisions. A stakeholder is either an individual, group or organization who is impacted by the outcome of a project.

Stakeholders are individuals, group, and institution important to the success of the project. Inmost of the situation a variety of stakeholders are likely to help some levels of interest or influence over a project, their expectation or what resources they commit. Influence refers to the power the stakeholders have offer a project such as their decision-making authority or their ability to influence project activities or stakeholders in a positive or negative way.

In expected list of stakeholders for the social development includes the following.

- Beneficiaries
- Customers
- Farmers
- Suppliers and Vendors
- communities
- investors
- Government
- People elected representatives. (Sarpanch, MLA).
- A district level officer (DC, CMO, DAO, DEO)
- Funding agencies / donor
- Project Volunteers
- Other project partners / NGOs.

Why to do stakeholder analysis?

Stakeholders' analysis help implementing agencies and partner, think carefully about who is important to your project and their level of interest or influence. From this analysis we can decide how shared this should be informed in stakeholder analysis helps.

a) Create and enhance right relationship among stakeholder & encourage corporation within a project.

b) Engage stakeholder appropriately so that they can participate meaningfully in all stages.

c) Reveal assumption and potential risks within the project.

How to do stakeholder analysis?

Project management team performed the comprehensive stakeholder analysis depending upon this needs to experience the project staff will be come at putting together at analysis i.e. appropriately solved. The project team makes understand the following step.

1. Identify the main purpose of analysis

What will your design team member to define the purpose of stakeholder analysis.

One reason might be to make sure you have identify the most important stakeholder that ought to be involved in assessment and analysis activity. Later on, descent making for project objective and strategy.

2. Identify stakeholders for the project

After established the purpose you can brainstorm a list of key stakeholder for your project at various level – local, regional and national. Be as specific as possible for eg; name individual rather than references, one you finish brainstorming & used the checked to help ensure that potential stakeholder will be identified.

3. Assess stakeholder interest, Influence and relationship

After the identification of stakeholder the rent slip is to list them in a table to further investigate their role relationship, interest, influence to participate in the project, resources required to involved stakeholder.

Assessment V/s Analysis

1. Assessment can be visualize as an abroad 'Horizontal process' while a wide number of issues are explored / (It shows the breath of the situation in

particular areas. Assessment is more open minded in term of question and issue to be studied). The issues, problems and opportunity uncaused by assessment are prioritize readiness for the analysis.

2. Analysis can be visualize as a deep vertical process where prioritize issues are product in-depth. It investigates the underlying cause and effect of specific problems or issues and involves reflection and examination.

Stakeholders of Agriculture

The major stakeholders of agriculture include farmers, government/s, buyers/ processors, vendors, traders, lenders, inputs/service providers and research institutions. The interests of farmers and the interests of government, to some extent, are already detailed earlier. The plans of the governments are also focused on support to agriculture with irrigation, power for irrigation, supply of inputs through public and private channels, access to desired support facilities/services, access to finance, support through programs/schemes, policies and regulations, investments required, infrastructure like rural roads, power, communication systems, irrigation systems, marketing infrastructure and such others. The processors or buyers would be interested in expected production, variants, and supply schedule, availability for their own processing and marketing or otherwise. The processors and buyers would also be interested in market opportunities for the products – both in raw and processed form. The inputs suppliers and services providers would be keen about demand for their products and services with focus on quality, quantity, and timing. Thus agricultural planning encompasses 'interests' of several stakeholders whereas this note is focused mostly on the farmers and to some functions of the government.

6

Implementation

Implementation

It is a process by which a act of predetermined activities is carried out in a planned manner, with a view to achieve certain establish objectives.

To implement a project means to carry out activities proposed with the aim to achieve project objectives and deliver results and outputs. It is a highly complex process and it involves a relationship among several system and variables.

Approaches

a) Traditional Approaches

b) Open Approaches

a) **Traditional:** It is characterized by mechanical view of planning and development in which implementation is a simply tool of (donors, plainness and politicians etc). Project beneficiaries are seen either as clients or employees.

b) **Open Approach:** It attempts to make people important thinking in terms of actor, planners and clients in a development situation and approach is needed which encourages a faster decreasing dependent and increasing same – and the open is suited to this.

Activities for strategic implementation

1. Planning for implementation
2. Developing support for implementation
3. Participation
4. Input resource planning
5. Project organization
6. Building backward and forward linkage.

Factors affecting implementation

A number of technical economical and factor affecting the implementation of a development programme knowledge about the nature, magnitude of the effect of each of this factor is necessary for development manager to be able to implement & managed the programme effectively and efficiently.

a) **Technical factor** – Aproject format can accommodate diverse activity from technical point of view. Project of NGO's may be as diverse as irrigation livestock development health education every project is aim at producing some output which may be an assets or a commodity or a function by which inputs are transferred to output. Production function involves is a technical in nature.

b) **Economical and financial factor** – Economic factor affecting a project a relevant from the point of view of the society as a whole whereas financial analysis takes the viewpoint of individual analysis.

   Financial analysis reveals the need for investment, credit, state to thane honoraria etc and other incentives for the successful implementation of the project. Economic analysis allows us to decide whether labour or other input to be used in the project should be remunerated at market prices.

c) **Commercial factor** – It affects the implementation of a project includes the marketing the output produced by the project and arrangement for the supply of input and credit needed to build and operate the project.

d) **Social–cultural factor** – Affecting the implementation of a project include the stratification of the project participants based on caste and religion, tradition, custom taboos. Mores, distribution of project benefit among the project. Impact an environment.

e) **Political factor** – NGO's have faced many challenges in implementing their project due to political output. Many NGO's left the area few fight fourth with politician and few compromise and acceptance of co-ordination with decentralization people elected representative have been dew place in emplaning and implementation.

f) **Managerial factor** – Managerial skills are necessary input for the optimal use of resources, resources mobilization information management monetary system, assessment of the needs of participation.

g) **People participation factor** – Many factor may motivate people to participate. It is necessary to find out factors and design specific participation. Some simple thumb rule for people participation are:

i) Create a human relationship

ii) Know the tradition and social customs

iii) Get us self a partner from local leaders

iv) Introduce programme gradually and adopt them to the ability of the target population.

v) Encourage and promote development leadership among the project employees & local people.

vi) Integration and co-ordination. Many Govt. and non-Govt. agencies are undertaking development programme at the grass root same time for the same area for same beneficiary. This is essential for optimum result. Otherwise it creates overlapping, duplication and wastage of scares resource.

Method for effective implementation

1. Bar charts or Gantt Charts: First developed by Henry L. Gantt

The bar chart is a pictorial representation showing various activities involved in a project. The chart has two co-ordinate axes one axes represent the activities and other axes represent the time required for completion of individual activities.

The axis represent activity involved in a project are drawn in the form of bar. In the completion of each activity. e.g: Construction of farmers' training centre.

Activity Time required		
1. Digging of foundation	-	3 weeks
2. Pouring foundation concrete	-	1 week
3. Construction of wall	-	10 weeks
4. Construction of roof slab	-	3 weeks
5. Fixing of doors & windows	-	1 week
6. Digging of well	-	6 weeks
7. Fastening of Construction	-	2 weeks
8. Electrification	-	1 week
		27 weeks

(All the activities can be depicted in a bar chart after identifying their logical sequence).

Network based scheduling – For agricultural or rural development project having large no. of activity the project scheduling become very complex. The network scheduling techniques can give away from time over-sum and cost time over-sum.

Project Evaluation and Review Technique/Critical Path Method (PERT/CPM)

PERT: PERT is used where the emphasis is on shortening and monitoring the project execution time without too much concern for its costs implications.

CMP: CMP is used where the emphasis is on optimising resources allocation and minimising overall cost for a given project execution time.

Network logic and terminology

The core of PERT-CPM is a network diagram which is also known as Arrow diagram. A network diagram consists of arrows and circles. Any project can be broken into a number of activities; the start and end of every activity can be recognised and are known as events in PERT terminology.

Activity

An activity is also known by other names such as job task, assignment, work etc. An activity consumes either time or resource or both, some examples of activities are.

A. Prepare project report

B. Prepare plan for water distribution

C. Prepare drawings

D. Big jack well

E. Erect pump

F. Construct high level tank

G Prepare field channels

Event

An event is also known as junction, node, stage, milestone etc., An event is a clearly definable movement of time which is the beginning or end of an activity or a number of activities. Examples of events are:

A. Preparation of project report started.

B. Drawing prepared

C. Boiler erected

D. Indents prepared

E. Materials received

F. Installation of machine started

Activity	Symbol	Activity Time	Symbolic Representation
Farmers training →	A	6 days	$\frac{A}{6}$
FPOs Training →	B	5 days	$\frac{B}{5}$

An activity is an identifiable job that has a beginning and end activity consumed resources like time human resource money and material like organizing health camp in villages, farmers training may be called activities. An activities is represented by straight arrow with circle both at the ends. The direction of the arrow indicates the direction of flow of project. The length of the arrow doesn't represent duration of activity. Circle in the beginning of arrow represent while circle placed at the end of the arrow shows the finishing point of activities

Network paths

From the network beginning event to any specific event there may be a number of network paths or chains of sequential events and activities. An activity begins in a start event or a tail event and ends in a completion event or head event. An event is represented by a numbered circle. IF an event represent a joint completion of more than one activity it is called 'merge events' and if an event represents a joint initiation of more than one activity a 'burst event'.

Simple network: Network is a graphical representation of a job plan showing the inter-relationships of the various activities. The Fig. Shows that a simple network consisting of 5 events (1 to 5) and its activities to be performed are designated as A to F. Activities A to F can be referred to as 1-2, 2-3, 3-4, etc.,

The logic of the network is that activities B and C cannot start until event 2 is reached, i.e. when activity A is completed B and C can be done concurrently. Activity E can start when event 3 is reached. i.e., when B is completed. Activity D connecting 3 and 4 shown by a broken arrow in the diagram is called a 'Dummy activity'. A dummy activity does not consume either time or resources but is used only to show inter-dependencies among activities.

For instance activity F depends not only on the completion of activity C but also on completion of activity B. It may be noted that E depends only on B while F depends on completion of activity C but also on completion of activity of both B and C. The project is completed when E and F are completed which can be done concurrently. Thus the network shows the interdependencies among activities and sequences in which they are to be performed.

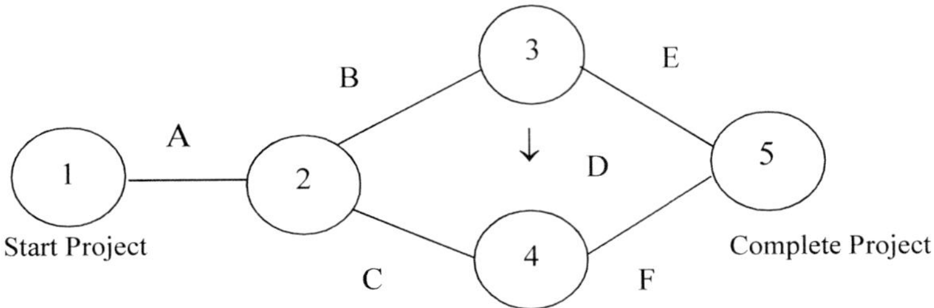

PERT (Project Evaluation and Review Technique) and CPM (Critical Path Method) are important network techniques useful in planning andcontrolling. These techniques are especially useful for planning; scheduling and implementing time bound projects involving performance of a variety of complex, diverse and interrelated activities. These techniques deals with time scheduling and resource allocation for these activities and aims at effective execution of projects within given time schedule and structure of costs.

Critical paths: Whatever may be the size and natures of a project whether it is building a house, land reclamation, lift irrigation, the project can be broken into a convenient number of activities. It has been observed that in any project 10 to 20 per cent of the total number of activities control the time required for the entire project. Any delay or speeding up of these activities will affect the final completion date. Others through essential to be performed do not affect the project time. In the simple network shown above there are three paths 1-2-3-5 and 1-2-4-5. If the time required for each of the activities laying in these paths is estimated the time for each path can be calculated. Time required for each path will be different (usually). The path taking the longest time is called the critical path. Evidently the time for the critical path is the minimum time within which the project can be completed. If it is required to complete the project in less time the duration of the activities laying in critical path will have to be shortened by employing more resources or by using improved technology. In a typical project there will be anywhere between 50 to 1000 activities depending upon the magnitude of the project and the extent to which activities are split up. In some cases there will be more than one critical path. Determination of critical path enables the management to concentrate on the activities on the critical path thus, releasing them from the need for controlling other activities.

CPM was developed in the year 1957, by Horgan R. Walks James E. Kelly

CPM is a network representation of logical sequence of activities having many parts starting from the initial & leading to last. If drawn of all activities that on a particular path are arranged it gives the duration of that path. The path with longest duration is called critical path and activities lie on the critical path are called critical activities. This is the critical path that set project.

The steps involved in using PERT/ CPM is as follows

- The project is divided into a number of clearly identifiable activities which are then arranged in a logical sequence.
- A network diagram is prepared to show the sequence of activities, the starting point and the termination point of the project.
- Time estimates are prepared for each activity. PERT requires the preparation of three time estimates – optimistic (or shortest time), pessimistic (or longest time) and most likely time. In CPM only one time estimate is prepared. In addition, CPM also requires making cost estimates for completion of project.
- The longest path in the network is identified as the critical path. It represents the sequence of those activities which are important for timely completion of the project and where no delays can be allowed without delaying the entire project.
- If required, the plan is modified so that execution and timely completion of project is under control. PERT and CPM are used extensively in areas like ship-building, construction projects, aircraft manufacture, etc.

Benefits of CPM

1. It provides a rapid and consistent common of information to increase progress and / or scheduled changes as reported by diverse organization.
2. Control cost and resource allocation.
3. Promoting team sprit and group judgment.
4. Aids better planning coordinating and controlling the budget.
5. It focuses management management attention on critical area and promotes management by acceptation.

7

Monitoring

MONITORING

Project monitoring is the process of keeping a close eye on the entire project management life cycle and ensuring project activities are on the right track. It is a continuous function that aims primary to provide functionaries with required feedback and early indication of progress or lack of their achievements of intended results.

Monitoring task is the actual performance or situation against what was planned or expected according to predetermined standard.

Monitoring generally involved collecting and analyzing data in implementation process, strategy and results and recommending corrective measures. Monitoring may be participatory and non-participatory.

Participatory monitoring

It is a process in which the project recipients become actively involved in implementation and regular checking on project progress or lack of it.

The step of Participatory monitoring

A Participating monitoring plan	Result Based Management			
BASIC C	1. Stakeholder concerns on type of result	Output	Outcome	Input
O M	2. Stakeholder concerns on indicators	Qualitative/Quantitative Indicators		
P O N	3. Definition of data collection & monitor the type of result	Type of Data when by whom where	Baseline Surveys, case studies, Documentary, informal meeting etc	Benchmark Targets
E N	4. Reporting requirement	Formats	Frequency	Distribution
T	5. Responsibility	Sub programme Implementers		
S	6. Responsibility	Crucial Monitoring and activities		
	7. Budget	For monitoring (1-5 %) of project cost)		

Indicators for monitoring

Result oriented monitory of development performance looking at results at the level of output outcomes and eventually impact.

Type of result (1)	What is measured (2)	Indicators (3)	Primary level of use (4)
1. Output	Effort or goods or services provide ed guaranteed by projects and programme.	Implementation of activities.	Project management.
2. Outcomes	Effectiveness access, usage and stakeholder satisfaction from goods and services profit and programme.	Use of output and sustained project benefit.	Programme management.
3. Impact	Improvement and development effectiveness / result of combined effect of a combination of outcome activities that improve condition at a national level.	Use of outcomes and sustain positive development changes.	Senior mgt.

Elements of implementation monitoring

Traditionally used for Project

a) Description of the problem and situation before the intervention

b) Benchmark for activity and immediate output

c) Data collection or input

d) Systematic reporting on provision of input.

e) Directly link to a series of intonation.

f) Design to provide information.

Elements of Implementation Monitoring: (Traditional)

Outcome monitoring (used for a range of intervention and strategies)

i) Baseline data describe the problem or situation before the intervention.

ii) Indicators for outcomes

iii) Data collection on output and how they contribute towards achievement

iv) More focus on perception of changes among stakeholder

v) Systematic reporting with more qualitative and quantitative information

vi) Done in conjunction with strategic partners

vii) Capture information with on success and failure achieving desire outcomes.

Monitoring consist of continuous collection of data on which its ongoing in the project and its effect in order to access progress towards the project objective .

Monitoring focuses on transforming input through activities into output. It is essential for management to make regular progress report on programme implementation that inputs are being made available as plan activities are taking place in line with work plan that outputs are being introduced.

National and International Monitoring System

Monitoring on a project depends on the attitude of the funding agencies, some prefer throw themselves, some through other and consultants sometimes beneficiaries are taken into believe in the monetary process.

Voluntary organization and people's organization invest act directly to increase and enhance people ability to make a national choice towards sustainable future. A voluntary organization working at international should collaborate in developing a system to monitor a plan and action to development donors, financial institution and Multi National Companies (MNC).

National Voluntary Organization must create similar system to monitor and make know the plans and actions of national govt. and co-operation.

The purpose to ensure that the people have access to be relevant information to access social and environmental intake and to take necessary impacts to protect interests.

Success ↔ Criteria

i) Impact

ii) Sustainability

iii) Contribution to capacity building.

Impact: Refers to results of a programme data access with reference to outcomes and impacts. In this content, outcomes and represent changes in a situation at a plan positive or negative, that a project or programme bring about.

Sustainability: It is the durability of positive programme or project result after the termination of the technical corporation channel through programme / project. Static sustainability refers to continuous flow of some benefits that we are set in motion by the completed programme or project to the same target group.

Dynamics sustainability: Refer to the use in or adoption of programme or project results to a different content or a changing environment by the original target group and other group.

Contribution to Capacity building: It is the extend to which a programme or project enables target group to be self resident and make win possible for government and for private sector to use positive expenses with the project in addressing broad developmental issues.

8

Evaluation

Evaluation

Project evaluation is a systematic and objective assessment of an ongoing or completed project. The aim is to determine the relevance and level of achievement of project objectives, development effectiveness, efficiency, impact and sustainability. It is a time bound exercise that attempt to access, systematically and objectively, the relevant / performance and success of a programme / project.

Evaluation is a complex process which seeks to identify the factors which are related to the performance and effectiveness of a programme / project in order to determine fixed to develop more effective project &programme in future.

1. Pre-Project Evaluation.
2. On-going Evaluation.
3. Ex-post Evaluation.

The concept of evaluation is an interrelated and an ongoing process includes monitoring. Evaluation should be joined effort & implementer in increasing the effectiveness of their programme, but also all individual involves in designing and implementing the programme.

Project evaluation in most of the planning project which involves systematic objective comprehensive appraisal of development programme for project. It implies an assessment of a project act its operational techniques, economic financial and managerial efficiency. Project evaluation refers to the procedure of fact findings about the results of planned social action. Evaluation is an essential act to policy to assess its success or failure and to find out lines of improvement.

Basis of evaluation

Evaluation on the basis of benefits: A project is evaluated on the basis of benefit acquiring. Benefits refers to addition of the flow of output acquiring from a project. There are:

a) Tangible benefits & Intangible benefits.

b) Real and None real benefits.

c) Direct and Indirect benefits.

On the basis of cost.

a) Real and Non Real cost.

b) Primary & Secondary cost.

Framework for monitoring and evaluation

1. **The selection of variables to be evaluated**

i) Critical variable feature important for the success of the project.

ii) Capable of being influenced by subsequent management decision.

iii) Uncertain performance in the given situation.

iv) Requiring a lower cost of evaluation.

2. **Alternative Evaluation Designs**

i) **Designs-One:** Case study with the measurement and no control group. This design only compared the only actual with the plan performances.

ii) **Design two:** A case studies with two measurement and control group. One measurement is before implementation & is after implementation. This design requires qualitative analysis to help overcome problem of changes.

iii) **Design-Three:** A case studies with one measurement with one control room. Data are collected on similar control group before during and after implementation of the project. This is the best evaluation design and also the most costly.

3. **Data collection, processing and presentation**

There are basically four sources of data

i) Exiting statistics

ii) Project report

iii) Surveys (specific to evaluation requirement).

iv) Key in formats (opinion leader or people selected representatives) (Nokma).

Processing is should help to determine the amount of data collected. The form & content of presentation should be the need and capability of the uses evaluation report of action oriented, easily understandable with appropriate levels of management.

Fives indicators for designing and implementing a programme evaluation

1) Why evaluate?

i) There are two reasons- For funding agency present an analyses programme outcome.

ii) For the NGO's or public programme staff to identify what work and why work?

2. **What should be evaluated?**

Good evaluation flow at the process of the programme and productivity of the project.

3. **What does cumulative evaluation indicates**

While formative evaluation examine programme, evaluation programme outcomes or results.

- Cumulative- at the end of the programme.
- Formative- during of the programme.

4. **What are some important thing conducting evaluation?**

i) Project staff responsible for implementing the project should not be treated as object of the evaluation.

ii) For in depth information, the outside evaluate to establish good report with staff that will facilitate the purpose of study.

iii) Participatory evaluation technique should be adopted to have a view of opinion leader. (Village head man and beneficiaries).

5. **What should the final report of evaluation include?**

i) A final report usually combine and integrates the formative and cumulative data. It analyses the relationship between what happen during the programme opinion and the programme outcome. A good evaluation report includes brief description of the programme.

ii) Narrative summary of the activities (problem and success).

iii) Results of formative and cumulative evaluation.

iv) Analyses of factor which help to achieve the goal.

v) Analyses of factor which hindered / obstacles to achieved the goals.

vi) Complete accommodation with alternatives.

Evaluating process of a project

→ Evaluation process includes four phases:

a) Preparation for evaluation

b) Conduct of evaluation

c) Preparation of report and recommendation

d) Distribution of evaluation report

A. **Evaluation worksheet**

i) Development Objectives

ii) Immediate Objectives

iii) Output Objectives

iv) Activities Objectives

v) Inputs

B. **Conduct of evaluation**

i) Effectiveness-to what extent is the project achieve or is lightly to achieve its objective.

ii) Efficiency-(to the expected project result continue to justify the cost).

iii) Relevance or significance- (Thus the project continues to make sense).

iv) Continuing validity of project design

v) Anticipated effect.

vi) Identification of alternatives.

vii) Casually (what factor affected project performance).

C. **Preparation of report & recommendations**

i) Overall conclusion (all project).

ii) Project revision (continuing project only).

iii) Decision and recommendation

iv) Identifying the lessons learned (Terminating and completed project).

D. **Distribution of evaluation report**

Evaluation Vs Monitoring

Monitoring	**V/s**	**Evaluation**
i) It keeps track of daily activities in a continuous basis.	i)	Periodically examine project effects or impacts (long range view)
ii) Accepts polices and rules	ii)	Question pertinence of policy & procedure.
iii) Looks at production of outputs	iii)	Examine progress toward object achievements.
iv) Focuses on transformation of input to output	iv)	Focuses on transformation of output to objectives.
v) Concentrate on planned project elements	v)	Assess planned elements and looks for unplanned changes.
vi) Reports on implementation progress	vi)	Checks on progress and seeks to identify lesson learned.

Reasons for failure of projects

1. Real problem of beneficiary not addressed
2. Socio cultural Value of the people not respected.
3. Less involvement of stakeholder in the designing of the project.
4. Objective not clearly and realistically defined.
5. Insufficient analysis of project idea before feasibility phase.
6. Project not embedded in a rational supporting framework.
7. Use of inappropriate technology.
8. Local management knowledge not taken into account.
9. Economic and financial aspect of project not elaborated.
10. Rest / no anticipated.
11. Lack of adequate inform and effective monitory.
12. Application of rules and procedures rather than managing project.
13. Insufficient clarity regarding the division of responsibility.
14. Through many managers in different many responsible for the same person.

Social cost benefit Analysis

In organization, investing in a project which to earn return from project. The return may be of commercial in nature maximization of profit or reduction in cost and it may give social benefit like employment opportunities. Benefit cost analysis (BCA) is a framework for evaluation of social cost and benefit of an investment project. This involves identifying measuring and comparing the private cost and negative externalities of a scheme with its private benefits and positive externalities, using money as a measure of value.

Step 1: Identity all cost and benefits using the opportunity cost (Alternative cost) (cost of rent alter foregone).

Step 2: Measure the benefit and cost using money as unit of account.

Step 3: Consider the livelihood of or benefit occurring.

Step 4: Take account of the timing of the cost and benefits.

Benefit Cost Analysis:-(BCA)

$$\text{B} - \text{C ratio} = \frac{\sum_{t=1}^{n} \frac{Bn}{(1+r)^n}}{\sum_{t-1}^{n} \frac{Cn}{(1+r)^n}}$$

Where

Bn	=	Benefit in each year
Cn	=	Cost in each year
n	=	No. of years
r	=	Interest rate (Discount rate)
t	=	time period

9

Project Appraisal Techniques

PROJECT APPRAISAL TECHNIQUES

Project appraisal technique gives details of cost and return of the proposed business. It involves investment of huge amount of capital. There are different project appraisal technique but the most common one is investment analysis or capital budgeting. With this, we can evaluate economic feasibility of different projects. For this, number of information are required like –

1. **Annual cash revenue:-** Different between the cash flow and cost flows of the project is called the annual cash revenue. This is required because the project cost and return are distributed is a number of ways.
2. **Total cost of the project:-** Total expenditure needed to be increased in the business.
3. **Terminal salvage value:-** It is also called as the junk value. It is the remaining amount of the depreciated assets of the project. In agricultural project, junk value is the taken as zero. Junk value is taken as market value that prevails in the determination of the projects.
4. **Discount rate or interest rate:-** It is also called the opportunity cost of the capital. It is the minimum rate of return which the amount invested in the project generate. If the project fails to generate minimum return, no capital is invested and alternative project is selected.

Classification of project appraisal techniques

1. **Undiscounted measures:-** These measures denote take into consideration the importance of time factor on cost and returns. It is used for ranking of projects different undiscounted projects are –
   a) **Ranking by inspection:** This method is based on the size of the cost and length of net cost revenue. If two projects have same cost and revenue, only differ is length of net cash revenue, i.e., one may give revenue for a longer period of time. Then the project giving revenue for a longer period of time is selected.

b) **Proceed per rupee of outlay:** (Proceeds – income/revenue; outlay – expenditure). The total proceeds generated by the project are divided by the total outlay. The value so obtained is used for ranking of projects.

c) **Average annual proceeds per rupee of outlay:** It is similar to proceeds per rupee of outlay, but average annual proceeds per rupee of outlay is calculated by dividing total proceeds by total number of years of the project. The value as obtained is further divided by total outlay. The value is used for ranking of projects.

d) **Payback period:** By this method, ranking is done by considering the length of time required to get back the investment made in the project.

$$P = \frac{I}{E}$$

Where,

P = Payback period

I = Investment made in the project

E = annual cash revenue

We, in our future transactions of goods and service make payment and receive the amounts at future dates. A person who deposit money in the bank receives the said amount after some specified date at agrees interest rate. The insurer pays the premium at specified regular intervals of time and receives the assumed sum in case of contingencies, accidents and so on. High paid professionals, expatriates, athletes, cine actors; managers offer their services for specified period of time on a given contract of salary and perks. In all such cases, we notice the concepts of time value of money and this forms the basis for formulating sound financial decisions.Let us consider the decisions problems like whether to have Rs. 1000 immediately or have it after a lapse of 3 years from now. Certainly the people prefer to have the money now, rather than having it at a future date. This is due to the fact that the people's want must be fulfilled now by having the desire goods and services. Another point to be worthy of mention here is that the value of the money over time falls and this trend continuous with time particularly during inflation periods. This means the rupee value in different time period as such is not comparable without, economic indices. To make sound and long term decisions, we require the money flow or payment at different time periods be reduce or standardized to common denominate, so that these flow can be compared. Future values of present sum and annuities should be worked out to help us in our financial decisions problems.

Capital Bugeting

Investment analysis is also called capital budgeting. The profitability of alternative agricultural project is determined through this technique. Four components are required for this analysis viz, net cash revenue from different projects, their cost, terminal or salvage value and discount rate to be used. The cash receipts minus cash expenses are the net cash revenue. The cost of investment is actual expenditure for its implementation. The terminal value of the project is equal to depreciation value of project and for simplicity junk value is assumed to be zero. The land values of the project should be estimate at its market value. Discount rate is opportunity cost of capital which represents the minimum rate of return for justifying the investment. If the proposed investment in the project fails to earn this minimum rate of interest, then capital should not be invested in the said project and alternative project must searched. If the capital is to be borrowed then the discount rate chosen should be higher than cost of borrowed capital.

Time value of money

Investment in agriculture is of two type's viz. operational investment and capital assets.Operational investment includes seed, feed, fertilizers etc. Where the time is not consider in profit maximization in operational investment because the expenditure and return are occur in the same year or even less.Capital assets includes land, machines, projects etc. Here time is consider in profit maximization because long time taken for getting return after expenditure. Therefore, there are certain instruments need to be analysis for getting profit maximization like recognition of time value of money, profitability and economic viability of capital investment.

Time value of money: Here consider two types of time like future and present.

1. **Future value of present money**: A rupee today is worth more than a rupee in future. This is primarily due to its opportunity cost, i.e., interest. Interest will be added to the principal over time and hence its value increases. Future value of present sum is an important concept in financial analysis and this is called compounding. In the compounding process, the interest is added to the principal at the end of each time period which, in turn, earns interest. The future value of present investment in the project is calculated by using the well-known formula of compound interest.

   $A = P(1 + i)^t$

Where,

A = Future value of the present sum invested in the project

P = Principal amount invested in the project

i = Interest rate in per cent and

t = Number of years.

Let us assume that investment made in an agricultural project is Rs. 10 core and that the expected rate of return from the project is 80 per cent. We are interested to know that would be the value of investment made after 40 years. This could be readily found using the equation.

Annuity: By definition annuity means a stream of payments or returns over time. The future value of annuity can be estimated using the following equation.

$$A = p\frac{(1+i)^t - 1}{i}$$

2. **Present value of future money:** The present value of future sum is the current value of investment to be received in the future. This present value is worked out through discounting process in which the future sum is discounted back to the present time to find out its current or present value. The rationale behind this process is that a sum to be received in future is somewhat less now, because of time difference assuming a positive interest rate. Discounting is the inverse procedure of compounding.
   - A present sum is compounded to know the future value and future sum is discounted to know the present value of future amount.

The present value of future money can be estimated using the following equation.

$$PW = \frac{P}{(1+i)^t} \text{ or } P\frac{1}{(1+i)^t}$$

Where,

PW = Present worth of future money

P = Money value in future

i = Rate of interest

t = Project life period in years.

The present value of annuity or stream of constant annual payment is found out using the following formula.

$$PW = P\frac{1-(1+i)^{-t}}{i}$$

Where,

PW = Present worth of future money

P = Money value in future

i = Rate of interest

t = Project life period in years.

Investment analysis is also called capital budgeting. Through the capital budgeting, the profitability of two or more alternative investment projects can be determined. Four components are required to analysis of investment they are

1. Net cash revenues from different projects (cash receipts less cash expenses)
2. Their costs (the cost of investment is the actual total expenditure)
3. Terminal or salvage value of investment (it is equal to the junk value for depreciable assets and the junk value is assumed to be zero for simplicity) and
4. Interest or discount rate to be used, here the discount rate is the opportunity cost of capital, which represents the minimum rate of return for justifying the investment. Opportunity is the next best alternative foregone cost.
   - If capital is to be borrowed for investment on the project, then the discount rate chosen for the economic analysis should be higher than the cost of borrowed capital.

Under the risk situation, the discount rate is to be equaled to the expected rate of return from alternative projects of equal risk.

Example

Compounding

Suppose a person decides to lend out Rs. 1000 for 5 years in a project at 5% rate of interest. So he is to be paid for the 5 years in the following manner-

Year	Amount at beginning of the year	One plus decimal value of interest	Amount at end of the year
1.	1000	1+0.05 = 1.05	1000 × 1.05 = 1050
2.	1050	1.05	1102
3.	1102	1.05	1157
4.	1157	1.05	1215
5.	1215	1.05	1276

Hence, for the loan amount of Rs. 1000 for 5 years at 5% annual compound interest at the end of the 5th year a total value of Rs.1276 would be done to be repaid.

Thus, as Rs. 1270 is the future value of Rs. 1000 invested at 5% for 5 years.

Discounting

It is the opposite off compounding. It is the present value of future payments (receipts) discounting at same rate.

Let us suppose we are to find out present worth of Rs. 1200 for 5 years in the future if the interest rate is assumed to be 8%. In order to answer this question we must divide the amount due to 1 + decimal value of interest (1.08) fir each project year as follows-

Year	Amount at end of the year	1 + decimal value of interest	Amount at beginning of the year
t_5	1200	1 + 0.08 = 1.08	1200÷1.08 = 1111
t_4	1111	1.08	1029
t_3	1029	1.08	953
t_2	953	1.08	882
t_1	882	1.08	817

Hence the process of finding out the present worth of future value is called discounting. Thus interest rate for discounting is called discount rate. So the only variation of compounding and discounting is the point of viewing.

The interest rate is used for compounding assuming a view point from here to future, whereas, discounting looks backwards (future to the present).

Present worth of stream of future income

An agricultural project often will return the same benefit in each of the several years, and we need to know the present worth of the future income stream to know how much we are satisfied in investing today to receive that income stream. To resolve this question, we will again need to know the rate of interest, the period of time we are taking about and of course the amount of income stream.

Suppose we are to receive the Rs. 6438 at the end of each year for 9 project years (t_1 to t_9). We can discount that income stream back to the present for each year using the discount factor at 15% of interest.

Year	Amount to be received	Discount factor at 15%DF = $1/(1+i)^n$	Present worth
t_1	6438	0.870	5598
t_2	6438	0.756	4868
t_3	6438	0.658	4233
t_4	6438	0.572	3681
t_5	6438	0.497	3201
t_6	6438	0.432	2783
t_7	6438	0.376	2420
t_8	6438	0.327	2105
t_9	6438	0.284	1830
Total	**57942**	**4.772**	**30719**

Thus, the present worth of Rs. 6438 received yearly for 9 years at a discount rate 15% is only Rs. 30719, i.e., the present worth is the sum of all present worth for all the years taken together. It will help us to estimate the level of the profitability in the business through investing some amount of money in the business.

Broadly there are two methods of project appraisal or analysis or evaluation namely undiscounted and discount techniques.

UNDISCOUNTED CASH FLOW MEASURES OF PROJECT APPRAISAL

Undiscounted measures are primitive, which often mislead in ranking of the project leads to wrong choices. Two important measures under this are 1. Pay back period 2. Rate of return method

Pay Back Period (PBP)

The length of time required to get the total investment made on the project is called pay back period. When two projects have same rate of return and the same account of risk, the decision may be taken on the basis of PBP.

P= I / E

Where,

P = Pack back period

I = Investment made on the project in Rs.

E = Annual net cash revenue in Rs.

Decision rules of PBP

1. Give highest ranking to the business investment / project which has shorter PBP, and follow ranking in descending order of PBP.
2. Give lowest ranking to the business investment which has longer PBP.
3. Generally the business investment / project with minimum or shorter PBP i. e. highest rank are accepted first for investment.

Limitation of PBP: It is inadequate to exercise the option among the alternatives, because it fails to consider very important points like consistency of running, timing of the proceeds, returns after the payback period and whether the cash-flow would be positive or negative in future.

Exercise

Calculate the pay back period of the following two projects and conclude your result. The initial investment in each project is Rs. 20,000/-

Year	Cash flows Project - A	Project -B
1.	5,000	4,000
2.	5,000	4,000
3.	5,000	4,000
4.	5,000	4,000
5.	5,000	4,000
6.	5,000	4,000

Project A = 20000/5000 = 4 years Project B = 20000/4000 = 5 years

Conclusion:Project 'A' with minimum or shorter PBP, so first rank is accepted for investment.

Rate of Return Method (ROR) or Rate on Investment Method (ROI)

ROR method expresses the profit generated by the investment as a percentage of the investment. In other words it is return per rupee of investment or ratio of earnings to investment. Its helps to know the profitability of an investment in the business or generation of returns per rupee of investment and to estimate the profits as percentage of investment.

Procedure

- Step 1 : Estimate the average investment which is equal to half of the original investment .Original investment can also be used
- Step 2 : Estimate the average annual net earnings over the life of the investment.
- Step 3 : Calculate the average rate of return by dividing the average annual net earnings or net profit by average investment.
- Step 4 : Average return per rupee of investment

  a. Divide the total returns by the number of years of investment to arrive at the average return per year

  b. Divide average return per year with original investment

$$\text{ROR} = \frac{\text{Net Pr ofit}}{\text{Original Investment}} \times 100$$

Or,

$$\frac{\text{Average annual profit}}{\text{Average Investment}} \times 100$$

Step 5 : Apply the decision rule as follows

a) Give highest rank to the investment in such a project / business whose ROR is highest.

b) Select the business /project whose investment yields highest ROR or having highest

DISCOUNTED MEASURES OF PROJECT APPRISAL

Cash flows are the yearly net benefits accrued from the project. If they are weighed or calculated by discount rate they become discounted cash flow. These discounted cash flow are the best estimate to decide on the worth of the projects. From the actual stream of gross benefits the capital invested plus other working cost are deducted to get the net present value. From that residual the return of capital as well as return to capital are computed. The residual is called the cash flow of the project.

The various discounted measure of project analysis are Net present value (NPV), Benefit cost Ratio (BCR), Internal Rate of Returns (IRR), N/K Ratio, Profitability Index etc.,

Let us know how these terms are computed.

NET PRESENT VALUE (NPV)

It is also called Net Present Worth (NPW) of the cash flows of project at a particular time period. The cash flow is actually the different cash inflows and cash outflows. The investment made in the agricultural projects is treated as cost of the project or simply it is cash outflows of the project. The various returns obtained from projects at different time periods is termed as cash flows or gross benefits of the projects. If the cash flows are discounted with appropriate discounted rate over the life span of the project, then it is called Net Present Worth (NPW) of the project. The selection criterion of the project depends upon positive value of NPV when discounted at the opportunity cost of capital. NPV is an absolute measure, not a relative measure.

Algebraically, it is worked out as

$$NPW = \sum_{t-1}^{n} \frac{Bt - Ct}{(1+i)^{t}}$$

where, B= Benefit received each year

C = Cost incurred each year

t = Time in years

n = Number in years of the project duration

i = Interest rate for discounting the cost/benefit

$$\text{NPW or NPV} = \left[\frac{P_1}{(1+i)^t} + \frac{P_2}{(1+i)^t} + \ldots\ldots\ldots\ldots + \frac{P_n}{(1+i)^{tn}} - Ct \right]$$

P_1 = Net cash flow in the first year;

t = time period:

C = initial cost of investment and

i = discount rate.

Utility of NPW

1. To determine the truth profitability of investment
2. It helps in ranking the alternative business investment proposal.
3. To choose among mutually exclusive projects that will maximize the benefits to investors.

Procedure of completing the NPV

Step 1: Forecast the cash outflow in the business.

Step 2: Forecast the cash-inflow from the investment.

Step 3: Estimate the net cash inflows by subtracting cash inflows from outflows.

Step 4: Identify the appropriate discount rate i.e. generally the opportunity cost of capital or required rate expected on investment.

Step 5: Work out the discount factors as the expected life period of the project.

Step 6: Calculate the net present value of the net cash flows (NPW) by multiplying the net cash flow with discount factors.

Step 7: Estimate the sum of the NPW for the life period of investment

Step 8: Apply the decision rule

a) Accept the investment in the business if NPW is positive or > 0

b) Reject the investment in the business if NPW is negative or < 0

c) If NPW = 0, (cash inflows = cash outflows) that means the cost of the investment has been fully recovered at the rate of discounting or the investment generates cash flows at a rate just equal to the opportunity cost of capital.

d) In the selection from a number of alternative investment proposal, the project would be arranged in descending order of their NPV's and given ranks i.e. first rank is given to the project with highest positive NPV and so on.

e) Select such projects or investment proposal with highest NPV's.

BENEFIT COST RATIO (BCR)

It is one of the discounted measures that are used to assess the credit-worthiness of the project. Here we compare the present worth of cost with present worth of cost with present worth of benefits. This ratio is obtained by dividing the sum of the present worth of benefit stream of the project with sum of the present worth of cost stream. The mathematical formula for working out this ratio is given as.

$$B-C\,Ratio = \frac{\sum_{t-1}^{n} \frac{Bt}{(1+i)^t}}{\sum_{t-1}^{n} \frac{Ct}{(1+i)^t}}$$

Where,

Bt = the benefit stream i.e. benefit of the project question in t^{th} year

Ct = the cost stream i.e. cost of the project in question in t^{th} year

t = 1 to t years i.e. life span of the project

i = the interest rate or discount rate at which funds are borrowed

Utility of BCR

1. Helps in selection of investment opportunities.
2. Ranking the project for implementation among various alternatives.

This ratio depends on the discount rate used. At higher 'I' value, the B-C ratio will be lesser. This ratio is more popular use I in resources project and private investment project. Generally, opportunity cost of capital is used to discount the benefits and cost of the project.

With a very little manipulation take the reciprocal of B-C ratio and subtract this from one. By doing so, we will come to know by what percentage the benefits could fall (B_{pf}), if B-C ratio is driven down to 1 from its present level, say 1.784

B_{pf} = 1- (1÷ B C Ratio) × 100

= 1- (1÷1.784) × 100

= 43.95%

So, in our example, if B-C ratio is driven to 1 from its value of 1.784, then benefits of the project would fall nearly by 44 percent.

Procedure of computing the BCR

Step 1: Estimate the cash outflows for the period of investment.

Step 2: Estimate the cash-inflows over the life period of the business.

Step 3: Identify the discount rate at which the costs and benefits are to be discounted.

Step 4: Work out the discount factors as the expected life period of the project.

Step 5: Calculate the net present value of the benefit of the benefits by multiplying the benefits of each year with the respective discount factor.

Step 6: Calculate the net present value of the benefits by multiplying the benefits of each year with the respective discount factor.

Step 7: Estimate the BCR by using the given formula.

Step 8: Apply the decision rule:

a) If benefit cost-ratio (BCR) of the project is more than one, then the investment in the project is credit worthy and hence the proposed agricultural project can be accepted for implementation. Similar decisions can be made even for deciding investment in different alternative agribusinesses.

b) It is better to work out BCR for alternative projects or business investment activities which are suitable to the given resources and environment and then select the project with the highest BCR among the alternatives for implementation.

c) Projects with BCR less then one should be dropped because they are unworthy projects.

d) If BCR = 1, (cash inflows=cash outflow) that means the cost of the investment has been fully recovered at the rate of discounting or the investment generates cash flows at a rate just equal to the opportunity cost of capital.

e) It is better to rank all the suitable alternative projects based on their magnitude of BCR. The first rank should be allotted to the project with the highest BCR.

INTERNAL RATE OF RETURNS (IRR)

Internal Rate of returns (IRR) is the discount rate which equates the present value of investment cash inflows i.e. Net Present Worth is equal to zero and the Benefit cost Ratio of project is equal to 1. In computation if IRR, time value of money is accounted. Method of working IRR provides knowledge of actual rate of interest from different project. So, known as marginal efficiency of capital or yield on investment. It is discount rate which must be found out with trial and error with some approximation. Here an arbitrary discount rate is assumed and its corresponding NPW is arrived. The positive NPW value indicates that IRR is still higher and next assumed arbitrary IRR value must be comparatively higher than the initial level. This process continues until NPW becomes negative.

It is the rate per rupee invested in agril. Project obtained over the life span of the project for example, if IRR of an irrigation project is 33%, it means the project would be given average annual return of Rs33 per Rs. 100 invested in irrigation project. This means at this IRR rate, the net present value of cash flow of the project (NPV) over its life span becomes zero.

Algebraically the IRR of the projects is worked out as

$$IRR = \sum_{t-1}^{n} \frac{Bt - Ct}{(1+i)^t} = 0$$

Where,

Bt = the benefit stream i.e. benefit of the project question in t^{th} year

Ct = the cost stream i.e. cost of the project in question in t^{th} year

t = 1 to t years i.e. life span of the project

i = the interest rate or discount rate at which funds are borrowed

The cost of credit (C) is the interest rate at which the project credit is borrowed. But in projects it is actually the social rate of discount (SRD). For project with long duration this discount rate generally less than 10%. The decision rule corresponding to IRR is, "Accept the project, if IRR is greater than cost of capital, (IRR > C)".

But in general, when

If NPV > 0, then IRR > C

If NPV = 0, then IRR = C

If NPV < 0, then IRR < C

The formula for interpolating the value of IRR lying between the two discounts rates (too high and too low) is given as

$$[IRR] = \begin{bmatrix} \text{Lower} \\ \text{Discount} \\ \text{Rate} \end{bmatrix} + \begin{bmatrix} \text{Difference} \\ \text{between the two} \\ \text{discount rates} \end{bmatrix} \times \left[\frac{\begin{array}{c}\text{Net present worth of the} \\ \text{cash flow at the lower} \\ \text{discount rate}\end{array}}{\begin{array}{c}\text{Absolute difference between net} \\ \text{present worths of the} \\ \text{cash flow at the two discount rates}\end{array}} \right]$$

Utility of IRR computation

1. Useful in identifying the discount rate by giving time adjusted value consideration.
2. To identify and select the rate of return which equates the present worth of cash inflows and outflows of an investment.
3. To know the average earning power of money used in the business.
4. To know the marginal efficiency of capital investment.
5. To select the business investment opportunities from alternative opportunities having rate of return greater than the opportunity.

Procedure of computation of IRR

Step 1: Estimate the total cost of investment each year.

Step 2: Estimate the total benefit from the business investment each year.

Step 3: Calculate the net benefits by deducting the costs from benefits of each year.

Step 4: Select a lower discount rate and a higher discount rate for discounting the net benefits.

Step 5: Identify the discount factors for each year at the lower and higher discount rates for the investment period.

Step 6: Calculate the present values at lower and higher discount rates separately by multiplying the year-wise net benefits with respective discount factors.

Step 7: Estimate the total of net present value at the two discount rates.

Step 8: Calculate the IRR using the above mentioned formula.

Step 9: Ally the decision rule to accept or reject the investment proposal in the business.

Decision rule

1. Accept the investment opportunity if IRR is greater than the market rate of interest or the opportunity cost of capital.
2. Reject the investment opportunity if IRR is less than the market rate of interest or the opportunity cost of capital.
3. When more than one investment opportunities are involved in the selection process:
   a) Arrange the investment opportunities in descending order of the value of IRR.
   b) Choose that set of opportunities for which IRR is greater than or equal to the market rate of interest, depending on the availability of funds.

Profitability Index: Profitability index (PI), also known as profit investment ratio (PIR) and value investment ratio (VIR), is the ratio of payoff to investment of a proposed project. It is a useful tool for ranking projects because it allows you to quantify the amount of value created per unit of investment. Under capital rationing, PI method is suitable because PI method indicates relative figure i.e. ratio instead of absolute figure.

The profitability index (PI), alternatively referred to as value investment ratio (VIR) or profit investment ratio (PIR), describes an index that represents the relationship between the costs and benefits of a proposed project. It is calculated as the ratio between the present value of future expected cash flows and the initial amount invested in the project. A higher PI means that a project will be considered more attractive.

$$\text{Profitability Index} = \frac{\text{PV of Future cash flow}}{\text{Initial Investment}}$$

Decision Rule

If PI > 1: Accept the project

If PI< 1: Reject the project

Profitability index helps in ranking investments and deciding the best investment that should be made. PI greater than one indicates that present value of future cash inflows from the investment is more than the initial investment, thereby indicating that it will earn profits.

PI of less than one indicates loss from the investment. PI equal to one means that there are no profits. Thus, profitability index helps investors in making decisions about whether or not to make a particular investment.

Merits of PI

1. It is superior to NPV
2. It gives due consideration to the time value of money and cost involved in the project.
3. PI techniques give better result in case of projects having different outlays.
4. In PI all cash flows are considered including working capital used and released, salvage value is also considered.
5. This method is considered best for wealth maximization of shareholders as it is based on cash inflow rather than accounting profit.
6. It considers total benefits arising out of project till the end of the project.

7. The discount rate applied for discounting the cash flows is actually the minimum required rate of return. This minimum rate of return incorporates both the pure return as well as the premium required to set-off the risk.

Demerits of PI

1. It is more difficult to understand.
2. It requires computation of required rate of return to be used as discount

Budgeting

Project budgeting is an estimation of the total costs for a project to be completed. Budgeting involves sharing limited resources between several group or for in a project environment. A project budget is the total sum of money allocated for the particular purpose of the project for a specific period of time. The goal of budget management is to control project costs within the approved budget and deliver the expected project goals. Budget analysis can serve the following purposes:

1. A plan for resource expenditure
2. A project selection criteria
3. A projection for project policy.
4. A basis for project control.
5. A performance measures.
6. Standardization of resource allocation.
7. An incentives for improvements.

Top Down Budgeting

Involves collecting data from upper level sources such as top & middle manages. The figure supplied by the manager may come from personal judgment, past experiences or past data or similar project activities. The cost estimates are passed to lower level manager who then break the estimate down to specific work component within the project.This estimate be given to like manager and lead worker to continue the process until individual activities chose are output.

Bottom up budgeting

In this method, elemental activities & their scheduled, description and labors skill requirements are used to construct the detail budget. Like workers familiar to specific activities are requested to provide cost estimate. Estimate & machine time. The estimates are then connected to the applicable. The estimate can be result through the intervention of senior manager, middle manager, project manager, account or standard cost consultant.

Steps to Create a Project Budget:

It is of utmost necessity to know how to plan, devise, create and then finally, manage a Project Budget.

Here is a set of steps to help with the same:

1. Create a task list to focus on requisite cost elements for the project.
2. Use historical data to research similar projects and the costing involved.
3. Estimate components based on project requirements, market research and find cost-effective and efficient alternatives.
4. Use market references to understand how competitors and seasoned companies manage budgetary control of their projects.
5. Take advice of experts to curb down doubts and obtain opinions to improve decision making.
6. Confirm accuracy of report with internal research and discussion with respective departmental heads.
7. Streamline the budget based on research and planning done in the above steps.
8. Test the budget on pilot projects for sample usage to understand efficiency and practicality.
9. Get approval from respective authorities.

Estimation of NPW for two Projects (Hypothetical)

Sericulture (one ha)						Mango Orchard (One ha)					
Year	**Costs (in Rs)**	**Returns (in Rs)**	**Net income (in Rs)**	**Discount factor at 12%**	**NPW (in Rs)**	**Year**	**Costs (in Rs)**	**Returns (in Rs)**	**Net income (in Rs)**	**Discount factor at 12%**	**NPW (in Rs)**
1.	38,900	-	-38,990	0.8929	-34,733.81	At the end of 6th	25,000	-	-25,000	0.507	-12,657
2.	9,239	28,475	19,236	0.7972	15,334.94	7th year	4,250	10,260	6,010	0.452	2,716.52
3.	10,575	32,550	21,957	0.7118	15,641.81	8th year	4,792	12,550	7,758	0.404	3,134.23
4.	11,952	35,610	23,658	0.6355	15,034.66	9th year	5,368	14,530	9,162	0.361	3,307.48
5.	12,858	39,802	26,944	0.5674	15,288.03	10th year	5,975	16,257	10,300	0.322	3,316.60
Total NPW					26,565.63	11th year	6,456	19,396	12,940	0.287	3,713.78
						12th year	7,187	21,470	14,283	0.257	3,670.73
										Total NPW	7,184.34

Benefit-cost Ratio Calculation for 2 Projects (Hypothetical)

Sericulture (One ha)						Mango Orchard (one ha)					
Year	Costs (in Rs)	Gross returns (in Rs)	Discount factor at 12%	Present worth of cost (in Rs)	Present worth of gross returns (in Rs)	Year	Costs (in Rs)	Gross returns (in Rs)	Discount factor at 12% (in Rs)	Present worth of cost returns (in Rs)	Present worth of gross
1.	38,900	-	0.8929	34,733.81	-	At the end of 6th	25,000	-	0.507	12,675.00	-
2.	9,239	28,475	0.7972	7,365.33	22,700.27	7th year	4,250	10,260	0.452	1,921.00	4,637.52
3.	10,575	32,550	0.7118	7,527.29	23,169.09	8th year	4,792	12,550	0.404	1,935.97	5,070.20
4.	11,952	35,610	0.6355	7,595.50	22,630.16	9th year	5,368	14,530	0.361	1,937.85	5,245.33
5.	12,858	39,802	0.5674	7,295.63	22,583.65	10th year	5,957	16,275	0.322	1,923.95	5,240.55
			Total	64,517.56	91,083.17	11th year	6,456	19,396	0.287	1,852.87	5,566.55
						12th year	7,187	21,470	0.257	1,847.06	5,517.79
						Total				24,093.70	31,278.04

$$\text{Benefit} - \text{Cost Ratio} = \frac{\text{Present worth of gross returns}}{\text{Present worth of costs}} = \frac{91,083.17}{64,517.56} = 1.41$$

$$\text{Benefit} - \text{Cost Ratio} = \frac{31.278.04}{24,093.70} = 1.30$$

Estimation of IRR Sericulture (Hypothetical)

Year	Costs (in Rs)	Gross income (in Rs)	Net income (in Rs)	Discount factor (40%)	Net present worth (in RS)	Discount factor (43%)	Net present worth (in Rs)
1.	38,900	-	-38,900	0.7143	-27,786.27	0.6993	-27,202.77
2.	9,239	28,475	19,236	0.5102	9,814.21	0.48902	9,406.4
3.	10,573	32,550	21,957	0.3644	8,007.69	0.3419	7,513.25
4.	11,952	35,610	23,658	0.2603	6,158.17	0.2391	5,656.62
5.	12,858	39,802	26,944	0.1859	5,008.89	0.1672	4,505.04
Total			52,913		1,202.69		-121.46

$$\text{RR} = 40 + 3\left(\frac{1202.69}{1,202.69 + 121.46}\right)$$

$$= 40 + 3\ (0.9083)$$

$$= 40 = 2.7249$$

$$= 42.7249$$

Estimation of IRR for Mango Orchard (One Hectare) (Hypothetical)

Year	Costs (in Rs)	Gross income (in Rs)	Net income (in Rs)	Discount factor (25%)	Net present worth (in RS)	Discount factor (30%)	Net present worth (in Rs)
End of 6th year	25,000	-	-25,000	0.262	-6,550	0.207	-5,175
7th year	4,250	10,260	6,010	0.21	1,262.01	0.159	955,59
8th year	4,792	12,550	7,758	0.168	1,303.30	0.123	954.23
9th year	5,368	14,530	9,162	0.134	1,227.71	0.094	861.23
10th year	5,957	16,275	10,300	0.107	1,102.10	0.073	751.90
11th year	6,456	19,396	12,940	0.086	1,112.84	0.056	724.64
12th year	7,187	21,470	14,283	0.069	985.53	0.043	614.17
Total			35,453		443.49		-313.24

$$\text{IRR} = 25 + 5\left(\frac{433.49}{433.49 + 313.24}\right)$$

$= 25 + 5\ (0.586)$

$= 25 + 2.93$

$=27.93$

Estimation of Profitability Index

Original amount invested in a Project= *Rs 60,000*

Year	Cash flow (in Rs)	Discounting factor (12%)	Net Present Worth (in Rs)
1.	14,500	0.8929	12,947
2.	14,900	0.7972	11,878
3.	16,600	0.7118	11,816
4.	18,700	0.6355	11,884
5.	19,000	0.5674	10,781
6.	20,000	0.5066	10,132
Total	1,03,700		69,438

$$\text{PI} = \frac{\text{Net present value of cash flows}}{\text{Original amount invested}}$$

$$= \frac{69,438}{60,000}$$

$$= 1.1573$$

Exercise

1. Estimate the BCR for the following two agribusiness and conclude the result

Sericulture Project

Year	Cost in Rs.	Returns in Rs	Discount factor @ 12%	Present worth of cost in Rs.	Present worth of benefit in Rs
1.	38900		$1/(1+0.12)^{1}$=0.8928		
2.	9239	28475	$1/(1+0.12)^{2}$=		
3.	10575	32550	$1/(1+0.12)^{3}$=		
4.	11953	35610	$1/(1+0.12)^{4}$=		
5.	12858	39802	$1/(1+0.12)^{5}$=		
Total NPW					

Mango project

Year	Cost in Rs	Returns in Rs	Discount factor @ 12	Present worth of cost in Rs.	Present worth of benefit in Rs.
End of 6th year	25000		$1/(1=0.12)^6=0.8928$		
End of 7th year	4250	10260	$1/(1=0.12)^7=$		
End of 8th year	4792	12550	$1/(1=0.12)^8=$		
End of 9th year	5368	14530	$1/(1=0.12)^9=$		
End of 10th year	5975	16257	$1/(1=0.12)^{10}=$		
End of 11th year	6456	19396	$1/(1=0.12)^{11}=$		
End of 12th year	7187	21470	$1/(1=0.12)^{12}=$		
Total NPW					

2. Work out the BCR of the following agribusiness enterprise.
(1 unit = 1 crore). The discount rate is 12.50%

Year	Incremental project costs			Incremental project benefits
	Capital expenditure in Rs.	Operating and maintenance cost in Rs.	Production cost in Rs.	
1.	2.8	0	0	0
2.	10.25	0	0	0
3.	11.46	0	0	0
4.	8.37	0	0	0
5.	4.21	0	0	0
6.	2.09	0	0	0
7.	0	1.28	0.86	6.44
8.	0	1.28	1.23	12.32
9.	0	1.28	2.47	16.74
10.	0	1.28	1.63	15.45
11 to 40.	0	1.28	1.63	15.87

3. Find out the BCR with 11.50% discount rate for the following two projects and suggest the better and feasible project with proper conclusions.

4. Estimate the IRR for the following two agribusiness projects and conclude the result.

Vermi-compost Project			Jatropa Bio-diesel Project		
Year	Cost (Rs.)	Returns (Rs.)	Year	Cost (Rs.)	Returns (Rs.)
1.	18,500		End of 5th year	35,800	
2.	6,300	16,500	End of 6th year	4,600	12,800
3.	7,400	20,600	End of 7th year	4,800	14,300
4.	8,000	22,000	End of 8th year	5,150	15,800
5.	8,300	23,800	End of 9th year	5,250	17,100
6.	8,450	25,100	End of 10th year	5,750	18,600
7.	8,100	25,850	End of 11th year	6,250	19,800

Value addition Project

Year	Cost in Rs.	Returns in Rs.	Net income in Rs.	Discount factor @ 49%	NPW	Discount factor @ 43%	NPW
1.	38900			$1/(1+0.40)^1$=0.7142		$1/(1+0.43)^1$=0.6993	
2.	9239	28475					
3.	10575	32550					
4.	11953	35610					
5.	12858	39802					
Total NPW							

Agricultural Technology Transfer project

Year	Cost in Rs.	Returns in Rs.	Net income in Rs.	Discount factor @ 25%	NPW	Discount factor@ 30%	NPW
End of 6th year	25000			$1/(1+0.25)^6$=		$1/(1+0.30)^6$=	
End of 7th year	4250	10260					
End of 8th year	4792	12550					
End of 9th year	5368	14530					
End of 10th year	5975	16257					
End of 11th year	6456	19396					
End of 12th year	7187	21470					
Total NPW							

5. Work out the IRR for the following agribusiness enterprise. (1 unit = 1 crore). The lower discount rate is 14.35% and the higher discount rate is 18.25%.

Year	Incremental project costs			Incremental project benefits
	Capital expenditure in Rs.	Operating and maintenance cost in Rs.	Production cost in Rs.	
1	2.84	0	0	0
2	10.25	0	0	0
3	11.46	0	0	0
4	8.37	0	0	0
5	4.21	0	0	0
6	2.09	0	0	0
7	0	1.28	0.86	6.44
8	0	1.28	1.23	12.32
9	0	1.28	2.47	16.74
10	0	1.28	1.63	15.45
11 to 40	0	1.28	1.63	15.87

6. Find out the **IRR(Internal Rate of Returns)** for the following two projects and suggest the better feasible project.

Vermicompost project			Jatropa Bio-Diesel Project		
Year	Costs (Rs.)	Returns (Rs.)	Year	Costs (Rs.)	Returns (Rs.)
1st year	38,900	Nil	End of 6th year	25,000	Nil
2nd year	9,239	28,475	End of 7th year	4,250	10,260
3rd year	10,575	32,550	End of 8th year	4,792	12,550
4th year	11,952	35,610	End of 9th year	5,368	14,530
5th year	12,858	39,802	End of 10th year	5,975	16,257
Discount Rates are 39.50% and 43.7% per annum			End of 11th year	6,456	19,396
			End of 12th year	7,187	21,470
			Discount rates are 26.25% and 29.75% per annum.		

10

Functions of Management

Management is the coordination and administration of tasks to achieve a goal. Management is a built-in function of business. It is essentially a decision-making process based on information and experience, in order to achieve desired goals and objectives. It involves creating an internal environment. It is the management which puts into use the various factors of production. Therefore, it is the responsibility of management to create such conditions which are conducive to maximum efforts so that people are able to perform their task efficiently and effectively. It includes ensuring availability of raw materials, determination of wages and salaries, formulation of rules & regulations etc. Therefore, we can say that good management includes both being effective and efficient. Management may in short be called a science of decision-making or a science of choice. A farmer has to make judicious decisions on the use of scarce resources, having alternative uses to obtain the maximum profit and family satisfaction on a continuous basis from the farm as a whole. In other words, management seeks to help the farmer in deciding problems like what to produce, how much to produce and when to buy and sell and in organization and managerial problems relating to these decisions.

Management is a process of planning, decision making, organizing, leading, motivation and controlling the human resources, financial, physical, and information resources of an organization to reach its goals efficiently and effectively.

- According to Kreitner, "Management is a problem-solving process of effectively achieving organizational objectives through the efficient use of scarce resources in a changing environment"
- According to F.W. Taylor, "Management is an art of knowing what to do when to do and see that it is done in the best and cheapest way"
- According to Harold Koontz, "Management is an art of getting things done through and with the people in formally organized groups". It is an art of creating an environment in which people can perform and individuals and can co-operate towards attainment of group goals.

- **Management** is the process of designing and maintaining an environment in which individuals, working together in groups, efficiently accomplish selected aims.
- Management is carrying out the managerial functions of planning, organizing, staffing, leading, and controlling.

Features of Management

Management is an activity concerned with guiding human and physical resources such that organizational goals can be achieved. Nature of management can be highlighted as:

1. Management is Goal-Oriented: The success of any management activity is accessed by its achievement of the predetermined goals or objective. Management is a purposeful activity. It is a tool which helps use of human & physical resources to fulfill the pre-determined goals. For example, the goal of an enterprise is maximum consumer satisfaction by producing quality goods and at reasonable prices. This can be achieved by employing efficient persons and making better use of scarce resources.
2. Management integrates Human, Physical and Financial Resources: In an organization, human beings work with non-human resources like machines. Materials, financial assets, buildings etc. Management integrates human efforts to those resources. It brings harmony among the human, physical and financial resources.
3. Management is Continuous: Management is an ongoing process. It involves continuous handling of problems and issues. It is concerned with identifying the problem and taking appropriate steps to solve it. e.g. the target of a company is maximum production. For achieving this target various policies have to be framed but this is not the end. Marketing and Advertising is also to be done. For this policies have to be again framed. Hence this is an ongoing process.
4. Management is all Pervasive: Management is required in all types of organizations whether it is political, social, cultural or business because it helps and directs various efforts towards a definite purpose. Thus clubs, hospitals, political parties, colleges, hospitals, business firms all require management. Whenever more than one person is engaged in working for a common goal, management is necessary. Whether it is a small business firm which may be engaged in trading or a large firm like Tata Iron & Steel, management is required everywhere irrespective of size or type of activity.
5. Management is a Group Activity: Management is very much less concerned with individual's efforts. It is more concerned with groups. It involves the

use of group effort to achieve predetermined goalof management of ABC & Co. is good refers to a group of persons managing the enterprise.

The basic components or elements of management include

1. **Planning**
2. **Decision making**
3. **Organizing**
4. **Directing**
5. **Coordination**
6. **Controlling**

1. Planning

Planning is the beginning of all the other processes of managementorganizing, staffing, directing, communication and control. Planning is the most fundamental. It is deciding in advance what to do, how to do, when to do and who is to do it. Planning bridges the gap from "where we are" to "where we want to go" (Koontz and O'Donnell). Before a manager can tackle any of the other functions, he or she must first devise a plan. A plan is a blueprint for goal achievement that specifies the necessary resource allocations, schedules, tasks, and other actions. In the words of Theo Haimann, "Planning is the function that determines in advance what should be done. It consists of selecting the enterprise objectives, policies, programmes, procedures and other means of achieving their objectives.Planning is the process of deciding organizational objectives and charting out the methods and strategies for achieving these objectives.

Characteristics of Planning

- Planning: Planning is an intellectual exercise. It is concerned with thinking in a creating way as to how the existing combination of resources may be adjusted and adapted to match the emerging opportunities. Planning enables the management to make decisions regarding (a) what is to be done; (b) how it is to be done; (c) when it is no done; and by whom it is to be done.
- Planning and Forecasting: Forecasting describes what one expects to happen if no changes are made to escape that happening. Planning describes what one wants to happen.
- Accomplishment of Group Activity: Planning is essential to any goal directed activity. It enables people with divergent perceptions and motivations to work together to achieve common goals.
- Choice between Alternatives: Planning seeks to adjust and adapt the existing mix of resources to meet the emerging opportunities. The first choice to be

made by management is with regard to objectives of the business, i.e., profitability, growth, consumer satisfaction, man power development, prestige, and so on. The next choice is in respect of the strategy to be adopted to accomplish the objectives. Then comes the operational part, i.e. determining the time frame, assignment of tasks and other resources for the accomplishment of the objectives.

- Pervasiveness of Planning: Involvement of managers at levels is essential to the success of planning.
- Flexibility: Successful running of an organization involves matching of its resources with the emerging opportunities in the business environment.
- Integrated Process: Planning involves selection of achievable objectives, and formulation of simple and realistic policies, programmes, procedures etc., for the accomplishment of that objective. Effective planning takes care of the conflicting views and settles for a course of action that is in the maximum interest of the organization, besides being satisfying to the personnel involved.

Nature of planning

- It is the primary function of management
- It is beginning of the process of management.
- It is an intellectual process which requires a manager to think before acting.
- It must be flexible
- It is an all pervasive function
- Based on objectives and policies
- Selective process
- Directed towards the efficiency
- Concerns with the future activities
- Based on facts

Importance of planning

- Minimizes risk and uncertainty
- Leads to success
- Focuses attention on the organization's goals
- Facilitate control
- Minimize the cost of performance
- Improving the competitive strength

Steps in planning

The steps generally involved in planning are as follows:

- Establishing verifiable goals or set of goals to be achieved
- Establishing planning premises: This helps in assumptions about the future and supply information relating to the future such as production trends, costs, economic conditions, prices, marketing conditions, predicting about the future.
- Deciding the planning period i.e. length of the time required in achieving the objectives.
- Finding alternative courses of action:
- Evaluating and selecting a course of action
- Developing derivative plans
- Measuring and controlling the progress: controlling is a critical part of any plan. Mangers need to check the progress of their plans so that they can (a) take whatever remedial action is necessary to make the plan work, or (b) change the original plan if it is unrealistic.

The planning is an expensive exercise, both in terms of time and money.

1. Identification of the Opportunity or Problem: Planning must facilitate the organization to suit it to its environment. The constraints and opportunities provided by the environment may be in the form of government regulations, existing cultural norms, limited financial resources in the capital market, changing technology, production of goods and services as per customer preferences etc. Hence, correct identification of opportunity or problem is the first step of planning.
2. Collection and Analysis of Relevant Information: Effective planning depends on the quality, relevance and validity of the information on which it is based. The sources of information may be classified as external and internal. External source will include suppliers, customers, professional people, trade publications, newspapers, magazines, conferences, etc. Internal sources will comprise meetings, reports, and contacts with superiors, same ranks and subordinates.
3. Establishment of Objectives: Establishment of objectives points out the desired outcome that an organization may aim at stability of operations, growth, a higher rate of return, market leadership and so on.
4. Determination of Planning Premises or Limitations: Planning has to take into account numerous uncertainties in its environment. Important components of the internal environmental are (a) technology, (b) structural relationship and organization design, (c) employee attitude and morale; and

(d) managerial decision-making process. Internal environment is within the control of management which can appropriately adjust and adapt it to the requirements of the external environment. Uncertainties relating to the environment are beyond the control of management. These may be in respect of (a) fiscal policies of the government, (b) economic condition; (c) population trends; (d) consumer tastes and preferences; e) competitor's plans and activities; and f) personal practices. Only those factors which are to critically affect the enterprise plans should be identified and evaluated.

5. Examining alternative course of action: Often, there will be more than one action plan to achieve a desired objective. For example, if the objective is to maximize profits and there are no limits to increasing production, the objective can be achieved through, either tapping unexpected markets, or intensifying sale efforts in the existing markets, or increasing the price, or diversifying production. The number of alternative plans prepared by a manager would depend on this imagination, skill and experience.
6. Weighing Alternative Courses of Action: Evaluation of each alternative action-plan will have to be from different points of view, namely, (a) its effectiveness in contributing to the accomplishment of organizational objectives; (b) its ability to withstand the effects of environmental changes; and (c) its integration with ongoing action plans.
7. Selecting a Course: Whether the evaluation of various alternatives is directed by individual preferences and prejudices, or it is based on mathematical and statistical techniques, the course of action is to be optimum, or the best under the circumstances. Selection of the best course of action depends on resource availability, objectives, efficiency and economy.
8. Determining Secondary Plans: Secondary plans flow from the basic plan. These are meant to support and expedite the achievement of the basic plans. For example, once the basic plan is decided upon, a number of secondary plans dealing with purchase of raw materials and machines, hiring and training of workers and so on would have to be prepared to facilitate execution of the basic plan.
9. Providing for Future Evaluation: In order to ascertain if plans selected for the purposes are proceeding along right lines, it is necessary to devise a system for continuous evaluation of plan.

Types of Plans

- Plans commit individuals, departments, organizations, and the resources of each to specific actions for the future. Effectively designed organizational goals fit into a hierarchy so that the achievement of goals at low levels permits the attainment of high-level goals. This process is called a means-ends chain because low-level goals lead to accomplishment of high-level goals.

- Three major types of plans can help managers achieve their organization's goals: strategic, tactical, and operational. Operational plans lead to the achievement of tactical plans, which in turn lead to the attainment of strategic plans. In addition to these three types of plans, managers should also develop a contingency plan in case their original plans fail.
  - **Operational plans**: The specific results expected from departments, work groups, and individuals are the operational goals. These goals are precise and measurable. "Process 150 sales applications each week" or "Publish 20 books this quarter" are examples of operational goals. An operational plan is one that a manager uses to accomplish his or her job responsibilities. Supervisors, team leaders, and facilitators develop operational plans to support tactical plans (see the next section). Operational plans can be a single-use plan or an ongoing plan.
  - Single-use: It plans apply to activities that do not recur or repeat. A one-time occurrence, such as a special sales program, is a single-use plan because it deals with the who, what, where, how, and how much of an activity. A budget is also a single-use plan because it predicts sources and amounts of income and how much they are used for a specific project.
  - Programmes: Programmes are a complex of goals, policies, rules, task assignments, steps to be taken, resources to be employed, and other elements necessary to carry out a given course of action and are ordinarily supported by the necessary capital and operating budgets. Designing programme therefore truly requires the most rigorous application of systems thinking and action.
  - Budget: It is a plan of statement of expected results expressed in numerical terms. It may be referred as a "numberisedprogramme". The financial operating budget is called as a profit plan and may be expressed in terms of labopur-hours, units of products, machine-hours,etc., It is a fundamental planning instrument in many companies because it forces a degree of definiteness in planning.
  - Continuing or ongoing plans: These are usually made once and retain their value over a period of years while undergoing periodic revisions and updates. The following are examples of ongoing plans:
  - A policy provides a broad guideline for managers to follow when dealing with important areas of decision making. Policies are general statements

that explain how a manager should attempt to handle routine management responsibilities. Typical human resources policies, for example, address such matters as employee hiring, terminations, performance appraisals, pay increases, and discipline.

- A procedure is a set of step-by-step directions that explains how activities or tasks are to be carried out. Most organizations have procedures for purchasing supplies and equipment, for example.Thisprocedure usually begins with a supervisor completing a purchasing requisition. The requisition is then sent to the next level of management for approval. The approved requisition is forwarded to the purchasing department. Depending on the amount of the request, the purchasing department may place an order, or they may need to secure quotations and/or bids for several vendors before placing the order. By defining the steps to be taken and the order in which they are to be done, procedures provide a standardized way of responding to a repetitive problem. A rule is an explicit statement that tells an employee what he or she can and cannot do. Rules are "do" and "don't" statements put into place to promote the safety of employees and the uniform treatment and behavior of employees. For example, rules about tardiness and absenteeism permit supervisors to make discipline decisions rapidly and with a high degree of fairness

Tactical plans: A tactical plan is concerned with what the lower level units within each division must do, how they must do it, and who is in charge at each level. Tactics are the means needed to activate a strategy and make it work. Tactical plans are concerned with shorter time frames and narrower scopes than are strategic plans. These plans usually span one year or less because they are considered short-term goals. Long-term goals, on the other hand, can take several years or more to accomplish. Normally, it is the middle manager's responsibility to take the broad strategic plan and identify specific tactical actions.

Strategic plans: A strategic plan is an outline of steps designed with the goals of the entire organization as a whole in mind, rather than with the goals of specific divisions or departments. Strategic planning begins with an organization's mission. Strategic plans look ahead over the next two, three, five, or even more years to move the organization from where it currently is to where it wants to be. Requiring multilevel involvement, these plans demand harmony among all levels of management within the organization. Top-level management develops the directional objectives for the entire organization, while lower levels of management develop compatible objectives and plans to achieve them. Top management's strategic plan for the entire organization becomes the framework and sets dimensions for the lower level planning.

Contingency plans: Intelligent and successful management depends upon a constant pursuit of adaptation, flexibility, and mastery of changing conditions. Strong management requires a "keeping all options open" approach at all times — that's where contingency planning comes in. Contingency planning involves identifying alternative courses of action that can be implemented if and when the original plan proves inadequate because of changing circumstances. Keep in mind that events beyond a manager's control may cause even the most carefully prepared alternative future scenarios to go awry. Unexpected problems and events frequently occur. When they do, managers may need to change their plans. Anticipating change during the planning process is best in case things don't go as expected. Management can then develop alternatives to the existing plan and ready them for use when and if circumstances make these alternatives appropriate.

Forms of planning

Strategic planning involves deciding what the major goals of the entire organization will be and what policies will guide the organization in its pursuit of these goals.

Strategic planning	Tactical planning
It decides the major goals & policies of allocation of resources to achieve these goals.	It decides the detailed use of resources for achieving each goals.
It is done at higher levels of management	It is done at lower of management.
It is long term	It is short-term.
It is generally based on long-term forecasts about technology, political environment, etc. and is more uncertain.	It is generally based on the past performance of the organizations and is less uncertain.
It is less detailed because it is not involved with day to day operations of the organization.	It is more detailed because it is involved with day-to-day operations of the organization.

Standing plans

- **Policies:** These are the general guidelines for decision making with limits or boundaries of operation leads to achievement of objectives.
- **Procedures:** A detailed set of instructions for performing a sequence of actions involved in doing a piece of work.
- **Methods:** These are prescribed way in which one step of procedure is to be performed.
- **Rules:** These are detailed and recorded instructions that a specific action must or must not be performed in a given situation.

Single use plans

- **Programmes:** Definite steps in proper sequence which need to be taken to discharge a given task it includes policies, procedures and budgeting.
- **Budget:** Financial statement indicating detailed allocation of policy to be persuaded for the purpose of achieving objectives.

It is statement of income and expenditure.Any budget must be quantitative process.

Limitations of planning

A manager's plans are directed at achieving goals. But, a planning effort encounters the following limitations:

- Planning is an expensive and time-consuming process.
- It stifles the imitativeness of the manager
- Its flexibility cannot be maintained and rapidly changing situations
- It is sometimes based on inaccurate premises
- It sometimes faces peoples resistance

2. Decision making

A decision is a choice between two or more alternatives. Weighing and measuring of the consequences of each alternative can be done in three types of conditions; certainty, risk and uncertainty. The word 'decides' means to come to a conclusion or resolution as to what one is expected to do at some later time.

According to Manely H. Jones, "It is a solution selected after examining several alternatives chosen because the decider foresees that the course of action he selects will do more than the others to further his goals and will be accompanied by the fewest possible objectionable consequences"'.

Decision is a choice whereby a person comes to a conclusion about given circumstances/ situation. It represents a course of behaviour or action about what one is expected to do or not to do. Decision- making may, therefore, be defined as a selection of one course of action from two or more alternative courses of action. Thus, it involves a choice-making activity and the choice determines our action or inaction. Decision-making is an indispensable part of life. Innumerable decisions are taken by human beings in day-to-day life. In business undertakings, decisions are taken at every step. All managerial functions viz., planning, organizing, staffing, directing, coordinating and controlling are carried through decisions. Decision-making is thus the core of managerial activities in an organisation.

Definition

Decision-making is the selection based on some criteria from two or more possible alternatives. "-—George R.Terry

A decision can be defined as a course of action consciously chosen from available alternatives for the purpose of desired result —J.L. Massie

A decision is an act of choice, wherein an executive forms a conclusion about what must be done in a given situation. A decision represents a course of behaviour chosen from a number of possible alternatives. -—D.E. Mc. Farland

Steps in decision making

- Recognizing the problem
- Deciding priorities among the problems
- Diagnosing the problem
- Developing alternative solutions or courses of action
- Measuring and comparing the consequences of alternative solutions
- Converting the decision into effective action and follow up of action

3. Organizing

To organize a project is to provide it with everything useful for its functioning; personnel, raw materials, tools, capital. All these may be divided into two main sections and human and material organizations.Organization means a system with parts which work together, or system with parts dependent upon each other. Once managers have their plans in place, they need to organize the necessary resources to accomplish their goals. Organizing, the second of the universal management functions, is the process of establishing the orderly use of resources by assigning and coordinating tasks. The organizing process transforms plans into reality through the purposeful deployment of people and resources within a decision-making framework known as the organizational structure. According to Louis Allen, organization is a process of identifying and grouping the work to be performed, defining and delegating responsibility and authority, and establishing relationships for the purpose of enabling people to work most effectively together in accomplishing objectives.

Nature or Characteristics of an Organization

1. Division of Labour: It is the root of any organization structure. In order to improve the efficiency of any organization, the total efforts of persons who joined together for common purpose have to be divided into different functions. These functions are further divided into sub-functions each to be performed

by different persons. After the division of the total effort into functions and sub-functions, the next step is to group the activities on the basis of similarity of work. For example, in a manufacturing enterprise, its total activities may be divided and grouped under (a) Production, (b) Marketing, (c) Finance and (d) Personnel. Even within each of these groups or departments, one or more sub-departments or sections may be created to look after particular activities.

2. Co-ordination: An organization has to adopt suitable methods to ensure proper co-ordination of the different activities perform at various work spots. This implies that there must be proper relationship between: (a) an employee and his work, (b) one employee and another and (c) one department or sub-department and another.
3. Objectives: Objectives of a business cannot be accomplished without an organization; similarly an organization cannot exist for long without objectives and goals.
4. Authority: Responsibility Structure for successful management, positions of personnel are so ranked that each of them is subordinates to the one above it, and superior to the one below it. Management authority may be defined as the right to act, or to direct the actions of others.
5. Communication: For successful management, effective communication is vital because management is concerned with working with others, and unless there is proper understanding between people, it cannot be effective. The channels of communication may be formal, informal, downward, and upward to horizontal.

Process of organizing

It is the process of defining and grouping the activities of the project and establishing authority relationships among them. In performing the organizing function the project leader differentiates and integrates the activities of his organization. Differentiation means the process of departmentalization or segmentation of activities on the basis of homogeneity. Integration is the process of achieving unity of effort among various departments/segments.

The important steps in organization process are as follows

1. Determining the Activities to be performed: The first step in this process is to divide the total effort into a number of functions and subfunction each to be performed, preferably, by a single individual or a group of individuals. Thus, specialization is a guiding principle in the division of activities.

2. Assignment of Responsibilities: It involves selection of suitable persons to take charge of activities to be performed at each work point. Also, the tasks to be performed by each member or group should be clearly defined.
3. Delegation of Authority: Along with the assignment of duties, there should be proper delegation of authority. It would be unrealistic to expect an individual to perform his job well if he lacks the authority to secure performance from his subordinates.
4. Selecting Right: Men for Right Jobs Before assigning a particular task to an individual, his technical competence, interests, and aptitude for the job should be tested. If the individual concerned lacks the technical ability to do his job, he can not perform it to the best of his ability.
5. Providing Right Environment: It involves provision of physical means like machines, furniture, stationery etc. and generation of right atmosphere in which employees can perform their respective tasks.

Steps in Organizing

a. Consideration of objectives
b. Grouping of activities into departments
c. Deciding which departments will be key departments
d. Determining levels at which various types of decisions are to be made
e. Determining the span of management
f. Setting up a coordination mechanism

Importance of Organization

- Efficiency in Management: Planning, direction and control can have meaning only when these functions are undertaken within the frame work of a properly designed and balanced organization. Organization is an effective instrument for realizing the objectives of an enterprise.
- Instrument of All Round Development: A balanced organization helps an enterprise to grow and enter new lines of business. It can achieve the necessary momentum and adaptability to meet the various challenges posed by the environmental forces.
- Adoption of New Technology: In a rapidly advancing world, changes are bound to take place in the techniques of production, distribution and manpower management. An effective management can foresee such changes in environment which will involve rescheduling of activities as a new approach to delegation of authority and responsibility

- Aid to Initiative: For an organization to continue to remain effective, it is necessary that it encourages-initiative among its staff. Then alone, it can discover talents and creativity among its employees.

4. Directing

Directing is concerned with the initiation of organized action and stimulating people to work. It involves issuance of orders, instructions and leading and motivating the employees to execute them. Directing is the inter-personal aspect of management which deals directly with influencing, guiding, supervising and motivating the subordinates for the accomplishment of pre-determined objectives. Planning, organizing, staffing are merely preparations for doing the work but the work actually initiates through directing function. Directing can be defined as that function of management, which helps in guiding and leading people to work in such a manner so as to perform efficiently and effectively for the attainment of organizational objectives. Directing is the managerial function, which initiates organized action. It is one of the most important fundamental functions of management and is a part of every managerial action taken because the direction is primarily concerned towards various other function of management like leadership, motivation, and communication.

Directing means giving instructions and guiding people in doing work. In our daily life, we come across many situations like a hotel owner directing his employees to complete certain activities for organizing a function, a teacher directing his student to complete an assignment, a film director directing the artists about how they should act in the film etc. In all these situations, we can observe that directing is done to achieve some predetermined objective. In the context of management of an organization, directing refers to the process of instructing, guiding, counseling, motivating and leading people in the organization to achieve its objectives. The term direction means to show the way of performing the activities. It indicates an action. The organization moves from assembly of resources to utilization.

According to Koontz and O 'Donnel "Directing is interpersonal aspect of managing by which subordinates are led to understood and contribute effectively and efficiently to the attainment of enterprises objectives."

According to Urwick and Breach: "Directing is the guidance, the inspiration, the leadership of men and women that constitutes the real case of the responsibilities management."

According to Massie: "Directing concerns the total manner in which a manager influences the action of subordinates. It is the final actions of manager in getting other's act after all preparations have been completed."

Characteristics of Directing

The main characteristics of directing are discussed below:

- Directing initiates action: Directing is a key managerial function. A manager has to perform this function along with planning, organizing, staffing and controlling while discharging his duties in the organization. While other functions prepare a setting for action, directing initiates action in the organization.
- Directing takes place at every level of management: Every manager, from top executive to supervisor performs the function of directing. The directing takes place wherever superior – subordinate relations exist.
- Directing is a continuous process: Directing is a continuous activity. It takes place throughout the life of the organization irrespective of people occupying managerial positions. We can observe that in organizations like Infosys, Tata, BHEL, HLL and the managers may change but the directing process continues because without direction the organizational activities can not continue further.
- Directing flows from top to bottom: Directing is first initiated at top level and flows to the bottom through organizational hierarchy. It means that every manager can direct his immediate subordinate and take instructions from his immediate boss.

Importance of directing are presented as follows

- Initiates Action: It helps to initiate action by the people in the organization towards attainment of desired objectives. The employees start working only when they get instructions and directions from their superiors. It is the directing function which starts actual work to convert plans into results
- Integrates Employee's Efforts: All the activities of the organization are interrelated so it is necessary to coordinate all the activities. It integrates the activities of subordinates by supervision, guidance and counseling.
- Means of motivation: It motivates the subordinates to work efficiently and to contribute their maximum efforts towards the achievement of organizational goals.
- Facilitates change: Employees often resist changes due to fear of adverse effects on their employment and promotion. Directing facilitates adjustment in the organization to cope with changes in the environment
- Stability and balance in the organization: Managers while performing directing function instruct, guide, supervise and inspire their subordinates in a manner

that they are able to strike a balance between individual and organizational interests.

5. Coordination

Coordination is the orderly arrangement of individual and group efforts to provide unity of action in the pursuit of a common goal. In an organization, for example, the purchase department buys raw materials for production, the production department produces the goods, and the marketing department to procure orders and sells the products. All these departments must function in an integrated manner so that the organizational goal is duly achieved. Thus, coordination involves synchronization of different activities and efforts of the various units of an organization so that the planned objectives may be achieved with minimum conflict. So, Coordination is the management of interdependence in work situations. Coordination leads to blending the activities of different individuals and group of individuals for the achievement of certain objectives. In an enterprise which consists of number of departments, such as production, purchase, sales, finance etc. there is a need for all of them to work in synchronization and achieve the organizational objectives.

Definition

- According to Henri Fayol "Coordination harmonizes, synchronizes and unifies individual efforts for better action and for the achievement of the business objectives."
- According to Brech, "Coordination is balancing and keeping together the team by ensuring suitable allocation of tasks to the various members and seeing that the tasks are performed with the harmony among the members themselves."
- According to Charles Worth opined that, "Co-ordination is the integration of several parts into an orderly hole to achieve the purpose of understanding"

Characteristics of Coordination

- It is not a distinct/separate function but the very essence of management
- It is the basic responsibility of management
- It does not arise spontaneously or by force
- The heart of co-ordination is unity of action/purpose
- It is a dynamic process (continuous or an ongoing process)
- It is required in group efforts, not in individual effort
- It has a common purpose of getting organizational objectives accomplished.

- It is necessary to all levels of organization
- It is a system concept.

Objectives of Coordination

- Reconciliation of goals: Conflicts in organization arise because of differences between organizational goals and individual goals and the individualistic perception of goals and its realization. Coordination is the only means by which such conflicts can be avoided. Through personal contacts and better communication conflicts are minimized and unity of purpose is achieved. Commitment to organizational goal is also brought about.
- Total accomplishment of goals: Although individuals are firmly committed for the achievement of organizational goals individual contribution to work brings about total accomplishment which is in excess of the aggregate of the individual contribution. This is realized through the establishment of a reporting system and clear-cut spelling out of business objectives.
- Harmonious relationships: Another objective of coordination is to maintain harmonious relationship between individuals and the organization. Individuals derive satisfaction when their work performance brings about realization of the desired goal. This keeps their morale high. As the organization is structured with clear lines of authority and responsibility, conflict between line and staff personnel is minimized and better relationship is established. This not only helps in reducing labour turnover, but also enables workmen to stick to their jobs in the organization. Thus, coordination in an organization is expected to promote good personal relationships.
- Economy and efficiency: Coordination aims at bringing about economy and efficiency of operations through synchronization of activities and individual efforts whereby wastage of resources is minimized and there is saving of time and expense. Reduced rejection and minimum delays in execution lead to efficiency of operations in the organization.

The management of project is based on the principle of specialization or division of labour. With jobs specialized and divided among units, coordination becomes necessary. Coordination is the management of interdependence in work situations. It is the orderly synchronization or fitting together of the interdependent efforts of individuals in order to attain a common goal.

Requisites for excellent coordination

- Direct contact
- Early start
- Continuity

- Dynamism
- Clear cut objectives
- Simplified organization
- Clear definition of authority and responsibility
- Effective communication
- Effective leadership and supervision

Importance of Coordination

When a number of people are working to carry out a task, coordination is the only method of synchronization. Through coordination, duplication of work and excessive burden on a single department can be eliminated. The task of coordination is becoming increasingly complex and difficult. The need for coordination arises because of the following factors

1. Division of labour: When managers divide work into specialized function or departments, they at the same time create a need for the coordination for these activities. Greater the division of labour, greater the need for the coordination.
2. Growth in size: With the growth in size of an enterprise and large employment of people, the task of integrating the activities becomes more complicated. To achieve the desired results its important to harmonize individual goals with organization goals through coordination.
3. Interdependence of units: The need for coordination in an organization also arises because of the interdependence of various units. Greater the interdependence of the units, the greater the need for coordination. As the entire department have their own set of policies and procedures, but to achieve the organization goals the activities of various departments. have to be coordinated.
4. Growing specialization: Modern business has become increasingly complex as various functions are to be performed by specialists. Specialization brings about need for more coordination because of diversity of tasks to be performed.

Techniques of Coordination

The important techniques of coordination are as follows:

- Clearly defined goals: The goals of the organization must be defined clearly. The objectives of the organization must be understood by the individuals. Well aware people will be able to contribute in achieving organizational goals.

- Cooperation: When the individuals of the organization help each other voluntarily, coordination becomes easier. This is known as cooperation. Cooperation takes place in the organization by ensuring harmonious relations among the peoples. The informal relations speed up the activity of the cooperation. Therefore, this should be promoted in the organization.
- Harmonized policies and practices: Policies, procedure and rules serve as guidelines for decision- making in a constant manner. Only the harmonized policies and procedure are helpful to achieve the goals of the organization.
- Sound organizational structure: Sound organizational structure are also known as a technique of the coordination. This is due to well defined authorities and responsibilities. Such organization does not create any problem and almost dispute free.
- Managerial hierarchy: Managerial Hierarchy in the organization is also responsible for coordination. At each level superior coordinate the activities and efforts of subordinates by means of authority.
- Communication system: A good communication system promotes cooperation and mutual understandings among individuals and groups. This is helpful in effective coordination because every aspect regarding organization clear to each and every employee.
- Liaison officer: Liaison officers are appointed in the organization to enhance the coordination between different sections, units and groups. These officers iron out the barriers of smooth coordination.

6. Controlling

Controlling is seeing that actual performance is guided towards expected performance. It is quite clear from the example that all managers need to manage situations intelligently and take corrective action before any damage is done to the business. Controlling function of management comes to the rescue of a manager here. It not only helps in keeping a track on the progress of activities but also ensures that activities conform to the standards set in advance so that organizational goals are achieved. All other functions of management cannot be completed effectively without performance of the control function. It implies measurement of accomplishment against the standards and correction of deviation, if any, to ensure achievement of organizational goals. The efficient system of control helps to predict deviation before they actually occur. Controlling ensures that there is effective and efficient utilization of organizational resources so as to achieve the organizational goals. Controlling has two basic purposes i.e. a) If facilitates coordination b) It helps in planning.

Control may be defined as the process of analyzing whether actions are being taken as planned and taking corrective measures to make them conform to the plan of action. Control is the essence of good management. It is concerned with ascertaining that planning, organizing and directing functions result in attainment of organizational objectives. It can be defined as comparison of actual performance with the planned performance.

- According to Henri Fayol, "Control consists in verifying whether everything occurs in conformity with the plan adopted, instructions issued and principles established."
- According to Theo Haimann, "Controlling is the process of checking whether or not proper progress is being made towards the objectives and goals and acting if necessary, to correct any deviation."

The main objective of controlling is to bring to light the variations between the standards set and performance and then to take necessary steps to prevent the occurrence of such variations in future.Controlling is checking current performance against predetermined standards contained in the lans, with a view to ensure adequate progress and satisfactory performance.

Characteristics of Control

Control is a device or a procedure which keeps the manager informed as the-activities for' which he is responsible and which assures him that his plans and policies are being carried out according to schedule. The nature of control function will be clearly understood from the following characteristics of control:

1. Control is all pervasive function: Control is essential at all levels of organization. It is a follow-up action to the other management functions. Every manager performs the control function irrespective of his rank and nature of job. Control is the essential counterpart of planning. It is the control function which completes the management process.
2. Control is a continuous process: Control is an ongoing and dynamic function of management. It involves continuous review of performance and revision of standards of operations. As long as an organization exists, control continues to exist. It is amenable to change with the external environment. Therefore it is a highly flexible process.
3. Planning is the basis of control: Control can be exercised only with reference to and on the basis of plans. Effective control is not possible unless the management spells out clear objectives of the organization. In fact, measurement of performance requires certain standards which are laid down

under planning. Planning sets the course and control ensures that actual action conforms to the planned action.

4. Action is the essence of control: Control is an action-oriented process. A manager initiates action which guides the operation within the sphere of plans. In order to ' prevent a recurrence of deviations a manager modifies or improves the existing plans,
5. Control is a forward looking process: Control aims at future. Although past experience is the criteria for future standards, control is concerned with checking the current performance and providing guidelines for the future. Therefore, control is both backward-looking and forwardlooking. It looks at future through the eyes of past.
6. Delegation is the key to control: Effective control requires adequate delegation of authority. An executive can perform the control function properly if he enjoys the authority. to take remedial action and is to be held accountable for results.
7. Control allows the organization to cope with uncertainty: Control helps in regulating the uncertain events of the organization. It anticipates any shift in task and preferences of consumers and directs the organization to modify its process in order to meet the contingencies of the future.

Importance of Controlling

Control is an indispensable function of management. Without control the best of plans can go awry. A good control system helps an organization in the following ways:

- Basis of future action: Control provides the basis for future actions. It will reduce the chances of mistakes being repeated in future by suggesting preventive steps.
- Facilitates decision making: The process of control is complete only when corrective measures have been taken. This requires taking a right decision as to what type of follow up action is to be taken.
- Facilitates discipline and order: The existence of control system has a positive impact on the behavior of the employees. They are cautious while performing their duties as they know they are being observed by their superiors.
- Facilitates Coordination: Control helps in Coordination of the activities of various departments of the enterprise. It provides them unity of direction.
- Facilitates motivation: A control system is most effective when it motivates people to high performance. Since most people respond to a challenge, successfully meeting a tough standard may provide a greater sense of accomplishment.

- Effective plan Implementation: Controlling and planning are interdependent. Control is the only means to ensure that the plans are being implemented control points out short comings of not only planning but also other functions of management. Comparison can be done through various Performance report, Personal Observation.

11

Sample Questions and Answers

(Objectives on Agricultural Project Management)

I. Choose the correct answer

1. Project is __________
   a) Investment activity b) Create productive assets
   c) Realising benefits over time d) All of these
2. Pre-feasibility study is required to __________
   a) To determine a promising investment opportunity
   b) To reap social benefits
   c) In-depth investigation
   d) All of these
3. Successful project management has achieved the objectives __________
   a) Within time-frame and cost
   b) At the desired performance
   c) Utilizing the resources effectively and efficiently
   d) All of these
4. __________ is/are stakeholders for the social development projects.
   a) District collector b) Village Head
   c) Funding Agency d) All of these
5. __________ is / are the basic elements of management.
   a) Planning
   b) Decision making & Organizing
   c) Coordinating and controlling
   d) All of these
6. Project design components is/are __________
   a) Objectives b) Activities
   c) Inputs and outputs d) All of these

7. __________ cost is incurred in construction, maintenance and execution of the project.
   a) Project cost b) Primary cost
   c) Real cost d) Social cost
8. The law of demands states that __________ and the other things remaining the same.
   a) Quantity demand is directly proportional to its price
   b) Quantity demand is inversely proportional to its price
   c) Quantity demand not related to price
   d) Demand and price are not related
9. __________ is timely collection and analysis of data on the progress of a project.
   a) Analysis b) Implementation
   c) Monitoring d) Evaluation
10. __________ is primarily meant to assess economic feasibility of the projects.
    a) Pre-project evaluation b) Concurrent evaluation
    c) Ex-post evaluation d) All of these
11. Success of projects depends on the __________
    a) Impact b) Sustainability
    c) Contribution to capacity buildings d) All of the above
12. The project having the highest IRR above the __________ cost of capital is selected.
    a) Opportunity b) Real
    c) Nominal d) None of these
13 Project investment is profitable and profit is worth while, if __________
    a) NPV > 0 b) NPV < 0
    c) NPV = 0 d) None of these
14. __________ is the rate of discount which makes the net present value of the project zero.
    a) B-C Ratio b) NPW
    c) IRR d) Pay back period
15. __________ measures the returns or benefit per unit cost or investment.
    a) NPW b) B-C Ratio
    c) IRR d) Pay back period

16. The time lag between the investment and returns from the project is termed as __________
    a) Investment period b) Lag period
    c) Gestation period d) None of the above
17. The point at which total cost curve and total revenue curve intersect each other is called..
    a) Break even point (BEP) b) Equilibrium point
    c) Both (a) & (b) d) None of the above
18. The evaluation process when the project is in execution is known as __________
    a) First phase evaluation b) Ex-post evaluation
    c) Concurrent evaluation d) End term evaluation
19. Incremental income due to increase in productivity is termed as __________
    a) Tangible benefit b) Intangible benefit
    c) Real benefit d) No benefit
20. The process of evaluation after completion of the project __________
    a) Mid course evaluation b) Ex-post evaluation
    c) Concurrent evaluation d) None of the above
21. Which of the following is not a function of management?
    a) Planning b) Staffing
    c) Co-operation d) Controlling
22. Management is __________
    a) An art b) A science
    c) Both art and science d) Neither
23. Assembling project team and assigning their responsibilities are done during which phase of a project management?
    a) Initiation b) Planning
    c) Execution d) Closure
24. The basic nature of a project is a/an __________ one.
    a) Permanent b) Temporary
    c) a) or b) d) Both a)and (b)
25. A process that involves continuously improving and detailing a plan as more detail become available is termed as __________
    a) Project analysis b) Project enhancing
    c) Progressive deliberation d) Progressive elaboration

26. A program is usually a group of __________
    a) Plans
    b) People and work
    c) Related projects
    d) Unrelated projects
27. Which from the following statement(s) is/are NOT true?
    I. Projects have defined objectives
    II. Programs have a larger scope than projects
    III. The projects and programs in a portfolio must be directly related
    a) I only
    b) II only
    c) III only
    d) II and III only
28. Project performance consists of __________
    a) Time
    b) Cost
    c) Quality
    d) All of the above
29. Five dimensions that must be managed on a project __________
    a) Constraint, Quality, Cost, Schedule, Staff
    b) Features, Quality, Cost, Schedule, Staff
    c) Features, priority, Cost, Schedule, Staff
    d) Features, Quality, Cost, Schedule, customer
30. __________ can be defined as a specifically evolved work plan densed to achieve a Specific objective within a specific period of time
    a) Idea generation.
    b) Opportunity Scanning.
    c) Project.
    d) Strategy
31. __________ is used to accomplish the project economically in the minimum available time with limited resources
    a) Project Scheduling.
    b) Network Analysis.
    c) Budget Analysis.
    d) Critical Planning
32. __________ is a form of financing especially for funding high technology, high risk and Perceived high reward projects
    a) Fixed capital
    b) Current capital
    c) Seed capital
    d) Venture capital
33. __________ is not a main element of the project management process?
    a) Estimation.
    b) Schedule.
    c) Monitor
    d) Systems design

34. __________ is a set of activities which are networked in an order and aimed towards achieving the goals of a project.

a) Project b) Process

c) Project management d) Project cycle

35. Resources refers to __________

a) Manpower b) Machinery

c) Materials d) All of the above

36. Developing a technology is an example of __________

a) Process b) Project

c) Scope d) All of the above

37. The project life cycle consists of __________

a) Understanding the scope of the project

b) Objectives of the project

c) Formulation and planning various activities

d) All of the above

38. __________ is (are) the responsibility (ies) of the project manager.

a) Budgeting and cost control b) Allocating resources

c) Tracking project expenditure d) All of the above

39. Project design phase consist of __________

a) Input received b) Output received

c) Both a) and (b) d) None of the above

40. According to the Project Management Institute (PMI), project management is defined as "the application of knowledge, __________, __________, and techniques to project activities to meet the project requirements".

a) Skills, analysis b) Tools, analysis

c) Analysis, theories d) Skills, tools

41. __________ is the first step in project planning.

a) Establish the objectives and scope.

b) Determine the budget.

c) Select the team organizational model.

d) Determine project constraints.

42. Risk must be considered in the _____ phase and weighed against the potential benefit of the project's success in order to decide if the project should be chosen.

    a) Completion b) Initiation

    c) Execution d) Planning

43. There is _______ correlation between project complexity and project risk.

    a) An unknown b) A positive

    c) A general d) A negative

44. During the _________ of a project, the project manager focuses on developing the project infrastructure needed to execute the project and developing clarity around the project charter and scope.

    a) Completion b) Start-up

    c) Execution d) Evaluation

45. A project budget estimate that is developed with the least amount of knowledge is known as __________

    a) Rough order of magnitude (ROM) estimate

    b) Scope of work estimate

    c) Conceptual estimate

    d) Milestone schedule estimate

46. __________ is the first step in developing a risk management plan?

    a) Analyze the risks.

    b) Estimate the likelihood of the risks occurring.

    c) Identify potential project risks.

    d) Develop a risk mitigation plan.

47. __________ is the most critical aspect in developing a project plan that meets project specifications within the timeframe and at the lowest costs.

    a) Assessing risk management __________

    b) Providing documents that specifically comply with the quality standards in use

    c) Requiring commitment to quality by all the employees and business partners

    d) Developing a project execution plan that matches the complexity level of the project

48. Risk represents the likelihood that an event will happen during the life of the project that will negatively affect the __________ of project goals.

a) Scope creep
b) Achievement
c) Float
d) Rough order of magnitude

49. __________ is true about a project manager.

a) Project managers analyze work processes and explore opportunities
b) Project managers are focused on the long-term health of the organization.
c) Project managers are process focused.
d) Project managers are goal oriented

50. The method of incorporating change into project planning and execution processes is called the __________

a) Project logic diagram
b) Change management process
c) Milestone schedule
d) Critical path

51. BCR >1 indicates __________

a) Benefits are greater than the costs.
b) Costs are greater than the benefits.
c) Costs and benefits are well balanced.
d) Benefits are less than the costs

52. __________ establishes key dates throughout the life of a project that must be met for the project to finish on time.

a) Critical path
b) Ballpark estimate
c) Pert diagram
d) Milestone schedule

53. The PERT in project management means program evaluation and _____ technique.

a) Resource
b) Reconciliation
c) Reconsideration
d) Review

54. "Risk" is usually _______ as the project progresses.

a) Increased
b) Reduced
c) Remained same
d) Become negligible

55. Tangible benefits of project is __________
    a) National Integration b) Standard of living
    c) Higher yield level d) Income distribution
56. Present value of future money is estimated by use of __________
    a) $P/(1+i)^t$ b) $P/(1+t)^n$
    c) $P(1+i)^t$ d) $P/(1+1/i)^t$
57. Shadow prices of resources are used in __________
    a) Technical analysis b) Financial analysis
    c) Economic analysis d) All of theses
58. Project planning and budgeting are used __________
    a) Economic aspects b) Commercial aspect
    c) Financial aspects d) Technical aspect
59. Sensitivity analysis is used in __________
    a) Preparation of project b) Appraisal
    c) Implementation d) Forecasting/ Evaluation
60. __________ is a procedure of fact-finding about the progress of the project.
    a) Preparation of project b) Appraisal
    c) Implementation d) Evaluation
61. __________ are technological externalities and technological spill-over acquired to the society due to the presence of projects.
    a) Project cost b) Direct cost
    c) Social cost d) Indirect cost
62. Project costs are __________
    a) Value of the resources in maintaining and operating
    b) Value of goods and services
    c) Current market prices
    d) All of these
63. Social cost involves in __________
    a) Pollution problems b) Health hazards
    c) Both a & b d) None of these

64. __________ is the art & science of decision making and leadership.

a) Planning b) Management

c) Marketing d) None of the above

65. __________ is a determined course of action.

a) Management b) Policy

c) Plan d) Strategy

66. The process of establishing a time sequence for the project work is known as __________

a) Objective b) Schedules

c) Procedures d) Budgets

67. Present worth of future money is estimated by __________

a) Discounting b) Annualizing

c) Compounding d) None of the above

68. Future value of money investment is estimated by __________

a) Discounting b) Equating

c) Compounding d) None of the above

69. The method which calculates the time of recoup initial investment of agricultural project in the form of expected cash flows known as.....

a) B-C Ratio b) NPW

c) IRR d) Pay back period

70. The net initial investment is divided by uniform increase in future cashflows to calculate __________

a) Investment period b) Discounting period

c) Pay back period d) Earning period

Answers

1	**d**	2	**d**	3	**d**	4	**d**	5	**d**	6	**d**	7	**b**	8	**b**	9	**c**
10	**a**	11	**d**	12	**a**	13	**a**	14	**c**	15	**b**	16	**c**	17	**a**	18	**c**
19	**a**	20	**b**	21	**c**	22	**c**	23	**a**	24	**b**	25	**d**	26	**c**	27	**c**
28	**d**	29	**b**	30	**c**	31	**a**	32	**d**	33	**d**	34	**a**	35	**d**	36	**b**
37	**d**	38	**d**	39	**c**	40	**d**	41	**a**	42	**b**	43	**b**	44	**b**	45	**c**
46	**c**	47	**d**	48	**b**	49	**d**	50	**b**	51	**a**	52	**d**	53	**d**	54	**b**
55	**c**	56	**a**	57	**c**	58	c	59	**d**	60	**d**	61	**c**	62	**a**	63	**c**
64	**b**	65	**c**	66	**b**	67	**a**	68	**c**	69	**d**	70	**c**				

II. **State whether the following statement are True/ False.**

1. Project management is an art of controlling the cost, time, manpower involved in a project.
2. Feasibility is the last step in the project management.
3. A well-planned project requires pre-feasibility study to explore the opportunities.
4. Project is a bundle of activities designed to get the desired results.
5. A project should only be planned when it has feasibility and sustainability.
6. To start any project, NGOs must ensure the feasibility of social acceptability.
7. The target group in the project is intended beneficiaries.
8. Appraisal should take place before the implementation of the project.
9. Evaluation can be done several times during the life of a project.
10. The market demand curve will have a positive slope.
11. Target groups can be individuals who are not included to get project benefits.
12. Project evaluation is the procedure of facts finding about the results of planned social action.
13. B-C ratio (BCR) method is discounted measure for project worth.
14. Social Cost Benefit Analysis (SCBA) is also referred to as Economic analysis.
15. The present value of a future sum is worked out through discounting method.
16. Project success requires any feed back in mid course.
17. Environmental appraisal is an important part of project management.
18. PERT is emphasized on time.
19. Economic feasibility test of a project is to show the technical knowledge how to be used.
20. Project outputs are products that can be reasonably expected from good management of the inputs and activities.
21. Projects are cutting edges of development.
22. Project management is a rationally planned and organized effort to achieve a specific goal.
23. Objectives of a project should be clearly defined, measurable, and achievable.
24. The most important problem faced in the implementation phase of a project is a delay in execution.
25. Scope, cost, and schedule are some of the parameters used for project negotiation.
26. Feasibility analysis is the first stage in the process of project development.

27. The technical feasibility aspect of a project relates to the earning capacity of the project.
28. Break-even is a financial term to describe a business or project where the sales revenue is equal to total expenses.
29. Good project plans always do not allow for flexibility to adapt to changing circumstances.
30. Project execution and control is where most of the resources are applied/ expended on the project.
31. Project planning is a significant facet of project management.
32. Project leader role will require an individual with strong management skills.
33. The Network diagram, also referred to as the project graph, shows the activities and events of the project and their logical relationships.
34. Activities must be so defined that they are distinct, logically uniform tasks for which time and resource requirement can be estimated.
35. CPM analysis is on variations in activity times as a result of changes in resource assignments.
36. PERT and CPM analysis largely overlooks the cost aspect which is usually as important as the time aspect and sometimes even more.
37. Resource allocation refers to fixing the available resources to the prioritised project activities.
38. Payback period is defined as the length of time required to recover the original investment on the project through cash flows earned.
39. The average rate of return of a project is the discount rate that makes the net present value equal to zero.
40. The internal rate of return is also called the accounting rate of return.
41. The probability of project risk depends on the project life cycle.
42. The impact of risks is higher in the initial stages of a project life cycle.
43. Proper risk management can increase the productivity and efficiency of the project team.
44. A survey may be conducted to identify the specific information requirements of stakeholders related to the various decision area in a project.
45. Gantt charts is a tool for examining and scheduling more complex projects.
46. Resource allocation problem is concerned with scheduling activities in such a way so as to find the shortest project schedule.
47. The decision of project termination affects all the Stakeholders of the project and can put some negative impact on the organisation's growth.

48. Critical path networks highlight which tasks are critical for a project to stay on schedule and which can slide without affecting the completion date of the project.
49. Scheduling means the process of deciding how to arrange resources between varieties of possible activities and tasks
50. PERT-CPM is the network based technique for effective formulation of project.

Answers of True and False

1	**T**	2	**F**	3	**T**	4	**T**	5	**T**	6	**T**	7	**T**	8	**T**	9	**T**
10	**F**	11	**F**	12	**T**	13	**T**	14	**F**	15	**T**	16	**T**	17	**T**	18	**T**
19	**F**	20	**T**	21	**T**	22	**T**	23	**T**	24	**T**	25	**T**	26	**T**	27	**F**
28	**T**	29	**F**	30	**T**	31	**T**	32	**T**	33	**T**	34	**T**	35	**T**	36	**T**
37	**T**	38	**T**	39	**F**	40	**F**	41	**T**	42	**F**	43	**T**	44	**T**	45	**T**
46	**T**	47	**T**	48	**T**	49	**T**	50	**T**								

III. **Math the followings: Column A with Column B**

	A		B
1.	Funding Agency	a)	Direct cost ( )
2.	Social Cost	b)	Utilization of scarce resources ( )
3.	Intangible benefit	c)	Increase in productivity ( )
4.	Economic aspects	d)	Health Hazards ( )
5.	Tangible benefit	e)	ICAR ( )
6.	Formulation	f)	Ex-post evaluation ( )
7.	Project	g)	Assess economic feasibility ( )
8.	End Evaluation	h)	National Integration ( )
9.	Pre project evaluation	i)	Location or Site selection ( )
10.	Primary costs	j)	Investment activity ( )

Answer

1	**e**	2	**d**	3	**h**	4	**b**	5	**c**	6	**i**	7	**j**	8	**f**	9	**g**
10	**a**																

Selected References

Baker, B.N., Murphy, P.C. and Fisher, D. (1983). "Factors Affecting Project Success," in D.I. Cleland and W.R. King, eds., Project Management Handbook (New York: Van Nostrand Reinhold): 778-801.

Broadway, A.C. and Broadway, A.A. (2009). Agri-Business Management. Kalyani Publishers, Ludhiana

Cleland, D.I. (1983). "Project Stakeholder Management," in D.I. Cleland and W.R. King, eds., Project Management Hand-book(New York: Van Nostrand Reinhold): 275-301.

Davis, J.C. (1984). "The Accidental Profession," Project Management Journal, 15,3: 6.

e- Krishishiksha, Learning Portal on Agricultural Education, ICAR, New Delhi.

Einsiedel, A.A. (1987). "Profile of Effective Project Managers," Project Management Journal, 18, 5 : 51-56.

Goselin, T. (1993). "What to Do with Last-Minute Jobs," World Executive Digest. p. 70.

Grahm, R.J. (1992). "A Survival Guide for the Accidental project Manager," Proceedings of the Annual Project Management Institute Symposium (Drexel Hill, PA: Project Management Institute), pp. 355-361.

Grey, C.F. and E.W. Larson. (2005). Project Management: The Managerial Process. New York: McGraw-Hill/Irwin.

Hatai, L.D. (2016). Agricultural Marketing Management, New India Publishing Agency (NIPA), New Delhi.

Horner, (1993). "Review of 'Managing People for Project Success,'" International Journal of project Management, 11: 125-126.

Hughes, T.P. (1998). Rescuing Prometheus. New York, Pantheon.

IBBS, C.W. and Y.H. Kwak (1997). "Measuring Project Management's Return on Investment." PM Network.

IBBS, C.W. and Y.H. Kwak, (2000) "Assessing Project Management Maturity." Project Management Journal, March.

J. Davidson Frame, (1987). Managing Projects in Organizations (San Francisco: Jossey-Bass.

Jagannadham,C., Rao, D. and Virmani, S.M. (2006). Developing Winning Research Proposals in Agricultural Research, Hyderabad, NAARM.

Johl, S.S. and Kapoor, T.R. (2011). Fundamentals of Farm Business Management. Kalyani Publishers, Ludhiana

Kerlinger F. (1973). Foundations of Behavioural Research Rinehart Winetons.

Kerzner, H. (2003) Project Management: A Systems Approach to Planning, Scheduling, and Controlling, 8th ed. New Jersey: Wiley.

Lokanadhan, K. et al (2009). Agri-Business Management. New India Publishing Agency.

Pandey, D.P. (2008). Rural Project Management, New age International (P) Ltd Publisher, New Delhi.

Pandey, P.S., R.P. Newman, and R.R. Cavanagh. (2000). The six Sigma Way, New York: McGraw-Hill.

Panneerselvam, R. and Senthilkumar, P. (2010). Project Management, PHI learning.

Prasanna C., (1989). Financia lManagement Theory & Practice. New Delhi: Tata McGraw-Hill Publishing Company Limited.

Prasanna C., (1995). Project Planning Analysis Selection Implementation & Review. New Delhi: Tata McGraw-Hill Publishing Company Limited.

Project Management Institute. (2004). A Guide to the Project Management Body of Knowledge, 3rd ed. Newtown Square, PA: Project Management Institute.

Pyzdek, T. (2003). The Six Sigma Handbook, rev. Ed., New York: McGraw-Hill.

Reddy, S.S et al (2012). Agricultural Economics. Oxford & IBH Publishers, New Delhi.

Sandberg, J. (2007). "Rise of the False Deadline Means Truly Urgent Often Gets Done Late." Wall Street Journal.

Shenhar, A. J., O. Levy, and D. Dvir. (1997). "Mapping the Dimensions of Project Success." Project Management Journal.

Sun, M. (1984). "Weighing the Social Costs of Innovation. "Science".

Tripathi P.C. and Reddy P.N, (2002). Principles of Management. New Delhi: Tata McGraw-Hill Publishing Company Limited.

Case Studies

Case Study-1

Agricultural Sustainability Project

Agriculture is known for its multi-functionalities of providing employment, livelihood, food, nutritional and ecological security. Agriculture is both a way of life and principal means of livelihood to 65 per cent of Indian population. The technological led growth of agriculture in India is recognized as the corner stone of our national strategy for food security, conservation of natural resources, rural employment and development and poverty alleviation. The emerging expectations of food, nutrition, employment and environmental protection call for higher productivity per unit land, labour, water and other inputs in an equitable and sustainable manner. Excessive and unbalanced use of agro-chemicals has led to increased production costs and dependence on external inputs and energy, decline in soil productivity, contamination of surface and ground water, and adverse effects on human and animal health. Since India needs to attain the food grains production target of 337 million tones to feed an estimated population of 1200 million, it is essential to look for an eco-friendly and popularizing ecology based low cost input and economically sustainable farming system. As a result of emerging environmental and food safety problems many are advocating that we should develop system of agriculture that would work with nature rather than those that attempt to conquer nature or try to change/tame nature.

Sustainable agriculture has gained acceptance as a conceptual and technological approach for shaping farming system of the future. Meeting the needs of the present without compromising the ability of future generations to meet their own needs is the key principle behind the concept of sustainability. Sustainable agriculture is low input and regenerative, which makes better use of farm's internal resources through incorporation of natural processes into agricultural production and greater use of improved knowledge and practices. Sustainable agriculture as the successful management of resources for agriculture to satisfy changing human needs while maintaining or enhancing the quality of environment and conserving natural resources (Food and Agriculture Organisation, 1991).

The term sustainability inherently evokes a concept of preserving and nurturing overtime. Sustainability is the ability to achieve continuing economic prosperity while protecting the natural systems of the planet and providing a high quality of life for its people. Sustainable agriculture integrates three main goals-environmental health, economic profitability, and social equity. Agricultural Sustainability refers to "maintenance and enhancement of productivity on a long term basis and should use improved and friendly technologies for economic development of agricultural sector".

Dr. M.S. Swaminathan, the eminent agricultural scientists, identified 14 major dimensions of sustainable agriculture covering the social, economic, technological, political and environmental facets of sustainability. Success in promoting sustainable agriculture can be achieved on seven fronts. i.e., Crop diversification, Genetic diversity, Integrated nutrient management (INM), Integrated pest management (IPM), Sustainable water management, Post harvest technology, Sound extension programmes. The increasing population, massive industrialization, agricultural transformation, under development etc. are the major crucial factors that have resulted in massive exploitation of natural resources viz. land, water, wild life etc. Thus there is urgency for sustainable development, which means the programmes should be environmentally sound, economically viable and socially acceptable.

Agriculture is the mainstay of state's economy and sustenance of the life of the people. It contributed about 25.75 per cent of NSDP for the State and provided employment directly or indirectly to around 65 per cent of the total work force as per the 2011 census. Odisha is endowed with richest natural resources in India. Since the development of agriculture in the state has lagged behind due to several constraints, such as traditional methods of cultivation, lack of modern technology in farming, low productivity, inadequate capital formation and low investment, inadequate irrigation facilities, uneconomic size of holdings, widespread illiteracy among farmers, helpless victim of natural calamities, inefficient management of resources, poor performance of extension education and inadequate agricultural marketing facilities. Odisha is purposively selected as consisting of 30 districts and there is great inequality; improper management and over exploitation of natural resources and explosion of population have created a threat to ecological balance, economic aspects and social status in different districts of the state. The persistently increasing inequality has become a threat to successful development of sustainable agricultural in the state. Hence, some efforts should be made in Odisha to evaluate the existing status and promote sustainable agriculture in the interests of sustainability and long term well being of the people.

The present case study is based on the following specific objectives

1. To examine suitable method for Sustainable Livelihood Security Index (SLSI) for agricultural sustainability
2. To evaluate the relative agricultural sustainability in different districts of Odisha
3. To analyse the utility of Sustainable Livelihood Security Index for evaluating the relative agricultural sustainability
4. To suggest the policy implications of agricultural sustainability in Odisha

The SLSI methodology is actually a generalization of the relative approach underlying the Human Development Index developed by the United Nations Development Programme (UNDP, 1990). It is a cross sectional measure in evaluating the relative sustainability status of a given set of entities. The Sustainable Livelihood Security Index (SLSI) has been proposed by Swaminathan (1991) to serve both as an educational and policy making tool for evaluating the potential for sustainable development (SD). The concept of Sustainable Livelihood Security (SLS) is defined by Swaminathan, as livelihood options which are ecologically secure, economically efficient and socially equitable. The intimate conceptual, casual and operational linkages between SLS and other welfare goals like poverty alleviation, meeting basic needs for human development and quality of life justify SLSI as a legitimate for SDA. The analytical approach essential for operationalising SLS in the form of SLSI is identified by the following propositions of SDA. Firstly, three-dimensional conceptions of the SDA are: ecological security, economic efficiency and social equity both in intra and inter-regional contexts. Secondly, dynamic and contextual nature of SDA, sustainability evaluation needs to be relative rather than absolute both in time and space. Lastly, in an operational context, the multi- dimensional conception of SDA requires the SLSI to be a composite of three indices, i.e., Ecological Security Index (ESI), Economic Efficiency Index (EEI) and Social Equity Index (SEI), so that it can take stock of both the conflicts and synergy between ecological, economic and equity aspects of SDA.

For any study on sustainable agriculture, the question arises as to how agricultural sustainability can be assessed. To empirically estimate SLSI, a simple approach has been followed involving the selection of a set of variables or indicators having the ability to say something more relevant and substantial about the ecological, economic and equity aspects of Sustainable Development of Agriculture (SDA). Although many indicators have been developed, they do not cover all aspects of sustainability. Moreover, due to variation in biophysical and socioeconomic conditions, indicators used in one region are not necessarily applicable to other regions. For instance, total twelve variables have been selected to illustrate three dimensions of SDA. Ecological security is assessed based on four variables: Population density (per km^2), Proportion of geographical area under forest (%), Cropping intensity (%) and Livestock density (per km^2). Economic efficiency is reflected by the four candidate variables: yield rate of rice (q/ha.), per capita output of foodgrains (kg/annum), fertilizer consumption (kg/ha.) and per capita income (Rs.). Social equity is represented by the following four variables: Female literacy (%), Infant mortality rate, Rural road connectivity (Km) and Villages electrified (%).

Moreover, despite their variation and limitations, the selected variables do have a good capacity to reflect the picture of the overall ecological, economic and equity aspects of a district's agricultural systems. The variables actually selected to represent a given dimension and also reflect the other aspects of agricultural sustainability. The secondary sources of data and general information for the twelve candidate variables for all the districts of Odisha were obtained from Directorate of Economics and Statistics, Odisha, and Odisha Human Development Reports.

The findings of this study revealed that the values of ESI, EEI and SEI range varies from 0.68 to 0.141, 0.75 to 0.075 and 0.701 to 0.209 respectively. This shows that the agricultural systems of the all districts of Odisha display wide variations in their ecological and social equity aspects relative to their economic aspects. While the SLSI indicates a range from 0.586 to 0.181 and the SLSI* reflects a range from 0.621 to 0.212. The results indicate that there is significant variation between SLSI and SLSI*. The SLSI* ranking of various districts differ significantly from their ranking based on SLSI. As a result, the effect of the weighting procedure of SLSI* range deflates the better performance slightly but inflates the poor performance substantially. Such an equalizing or normalizing effect is favourable for districts with poor performance as inter-districts priority for investment in allocation of resources will be inversed to their ranking. The relatively narrower range of SLSI and SLSI* as compared to their component indices describes that the performance of districts is not consistent across the three aspects (ESI, EEI and SEI) of sustainable Development of Agriculture (SDA).

As analysed in the following section, the SLSI* ranking implies that the districts having the best conditions for sustainable development of agriculture is Baragarh followed by Jagatsinghpur and Jajpur. Similarly, the districts having the least desirable conditions for SDA is Nuapada followed by Deogarh and then Boudh. The SLSI* ranking appears to effectively identify the advanced and backward districts. It can be observed that the districts with the better SLSI* ranks are often described as advanced districts using other ecological, economic and social indicators. On the other hand, the districts with the lower SLSI ranks are generally known as backward districts i.e., the districts with poor conditions for sustainable development of agriculture over the reference time period. Hence, SLSI* reflects the picture of the overall performance of a district, its component indices indicate how the districts fares in the three dimensions of sustainability. It can be noted that Baragarh districts has the highest SLSI*, but in comparison of three indices its performance of ecological security index is not so good as its economic efficiency and social equity. On the other hand, in case of simple SLSI ranks Jagatsinghpur district has occupied the top position followed by Balasore and Cuttack. Overall, the coastal districts having the better performance of SLSI as compared to other parts of Odisha.

Consequently, the overall performance of the districts in terms of their both SLSI and SLSI*; only eight districts (about $1/4^{th}$) out of 30 districts in Odisha have an index of above 0.5, while thirteen districts have an SLSI* lower than 0.4. Moreover, many of the districts of coastal Odisha have better performance in agricultural sustainability in comparison to the districts of western Odisha as a whole.

Sustainable Agriculture (SA) became a dominant global concern for livelihood improvement and environmental stability. The future challenges and policy implications of the SLSI* approach has received increasing attention from the fact that it helps not only to establish inter-districts priority for the allocation of agricultural investment but also to priorities the activities and programmes relevant to each district for sustainable agricultural development. The districts with an SLSI* of less than 0.4 i.e., poor conditions for SDA, should receive the prim-importance in agricultural investment, policy making and successful implementation. In view of the inter-districts investment allocation, the SLSI* approach could also influencing recognition and policy guidance on the context specific activities and programmes to improve the overall agricultural sustainability of each district successfully. Many initiatives towards sustainable agricultural system, a better use of local resources and better management of the environment should be implemented. As a result, these experiences build on people's own knowledge, skill and their values, resources, culture and institutions and lead to the empowerment of farming communities. Hence, sustainable agriculture can be developed successfully in the state of Odisha.

Any given policy of SDA will influence among producers, consumers, agribusiness, traders, academicians, researchers, policy makers, input suppliers, food processors and others for the successful management of natural resources, biodiversity, food and nutrition security, ecosystem services and many other challenges on the way of supporting people living in rural areas of the state Odisha. Some efforts have been made in Odisha to promote sustainable agriculture in the interests of sustainability and long term well being of the people.

In this regard, the sustainable agriculture is being a pioneering endeavour aimed at making agriculture environmentally sound, economically viable and socially acceptable. The future programmes of SDA is a emerging challenge to develop new strategies and solutions to the multitude problems of agriculture and allied sectors associated with better sustainable development in the new millenniums.

Case Study-2

Agricultural Marketing Information System (AMIS) Project

Marketing Information System (MIS) is a process of gathering, processing, storing and using information to make better marketing decisions and to improve marketing exchange. A market information system is an important tool used by modern management to aid in problem solving and decision making.Farmers need information to aid them in planning their operations right from the time they plant the seeds until the produce passes the hands in the market. Agricultural marketing information helps the farmers in comparing the prices offered by different firms in different markets and also in the selection of alternative outlets available. The Agricultural Marketing Information System (AMIS) reduces business risks of farmers, sellers and traders. Improvement of agricultural market information services was necessary for domestic market efficiency and to integrate domestic agricultural market with regional and international market for sustainable development of agriculture sector and to ensure country's long run food security. There is need of proper dissemination of market intelligence and information through all possible means of communication for improving the marketing efficiency.

Meghalaya is a state of great diversity and inequality in many aspects of life and nature. It needs special attention for development of agricultural marketing information in the state. The study would help planners, policy makers, agriculturist, researchers and government bodies by highlighting the different aspects of agricultural marketing information system of Meghalaya.Realizing the deficiency of traditional market information and with the advent of information technology, the government developed the system to link the market data on agricultural commodities at the national level through AGMARKNET. Similarly, the Meghalaya government has developed the network within the state and uplinked to AGMARKNET. Day to day market trade information on agricultural commodities is collected at all important Agriculture Produce Market Committees (APMC) in the state.There are at present, two APMCs in the state of Meghalaya namely, Mawiong Regulated Market in Mylliem Block of East Khasi Hills and Garobadha Regulated Market in Selsella Block of West Garo Hills district.

Objectives of the case study

1. To find out the various sources of existing agricultural marketing information system in Meghalaya.
2. To study the pattern and extent of dissemination and utilization of existing agricultural marketing information by different stakeholders.
3. To identify the constraints in the agricultural marketing information system in the study area.
4. To suggest appropriate policy measures to stakeholders for implementation of agricultural marketing information system in Meghalaya.

To study the existing agricultural market information system (AMIS) and its dissemination, two regulated markets namely, Mawiong Regulated Market in Mylliem Block of East Khasi Hills and Garobadha Regulated Market in Selsella Block of West Garo Hills district were selected purposively for this study for easy accessibility of agricultural marketing information and so as to cover entire state. Further, the arrivals of agricultural commodities in these markets are highest. Arrivals and prices information for the same period was also collected from the Tura and Shillong markets as these markets were considered to be the reference markets for the State. To study the sources of agriculture market information and their utilization among the farmers and traders, 60 farmers from each selected market area and 20 traders from each regulated market were selected for the study based on random sampling technique. Post enumeration classification of farmers into 30 small (< 1 ha.), 20 medium (1-2 ha.) and 10 large (> 2 ha.) farmers was done based on the size of land holding.Besides, data from all the two selected market committees were collected to know the infrastructure facilities available in those markets with respect to collection and dissemination of agriculture market information. Thus the sample size consisted of 120 farmers, 40 traders and 2 market committees.

The primary data from sample farmers, traders & officials of regulated markets was collected by personal interview method by using pre-tested structured schedule prepared for the purpose. The data on area, production, arrivals, prices, exports, etc. were elicited from secondary sources. To understand the market information system for agricultural commodities, both tabular and econometric models were designed to analyze the data of the study. To find out the nature, extent, sources, utilization and expectations of market information system by farmers, traders and officials, tabular analysis with simple averages, percentages, etc, were computed. Interview method was developed to get complete and reliable information with the help of well structured schedule.

The study highlighted that the ultimate objectives of enhancing access and use of agricultural marketing information (AMI) is to improve the livelihood and empowerment of farmers in the study area of Meghalaya.

- **Pattern of awareness on agricultural market information**: The study revealed that in case of large category of sample farmers, the extent of awareness on arrivals, prices in local markets and other markets, quality / grade of produce required, post harvest handling of agricultural produce was found to be higher than small and medium size farmers. It was found that none of the small farmers aware of the area of crop cultivated, production and export & import of the agricultural produces.Poor awareness of farmers on available market information clearly highlights the need to create awareness on market led extension among the farmers through the agricultural extension agencies like the State Department of Agriculture, Krishi Vigyan Kendras so that the marketing information on agriculture commodities are incorporated in the extension services along with production aspects to the farmers. Generally, the degree of awareness of agricultural marketing information by traders was found to be considerably higher than different categories of sample farmers.
- **Sources of agricultural market information:** It was observed that the sources of agricultural market information at household level were radio, newspaper and television for small farmers. Radio formed the major sources of information for medium farmers at household level. The major sources of AMI at village level were neighbours followed by friends for small farmers. At the market level, commission agents were most predominant sources of AMI for all categories of farmers. It revealed that contacts in other markets over phone and fellow traders were the major sources of agricultural marketing information among traders in the study area.
- **Pattern of collection and documentation of AMI :**It was found that market arrivals and prices (maximum, minimum and modal) were the only two major types of market information documented and made available to the farmers and other intended beneficiaries in both Garobadha and Mawiong regulated markets. It was found that there was no documentation of information like area under cultivation, production, post harvest handlings of agricultural produce, value addition and pattern of packing in the selected markets. The arrivals and prices information were documented both in written form and electronic form at the main market of Mawiong and sub markets of Garobadha regulated market. Moreover, such arrivals, prices and quality aspects of information were documented on daily basis in Mawiong regulated market and weekly basis in Garobadha regulated market. Personal visit to the main market and sub market yard by the personnel and regular employees

was the major method of collection of market information on arrivals and prices in both the selected regulated markets. In general, the methodology adopted for generating arrivals and prices information was based on entire population recorded from the entry point and commission agent's transactions of the day in both the selected regulated market.

- **Pattern of dissemination of agricultural market information:**The market information was disseminated through notice boards, announcements, telephone, internet, AIR, personal contacts and newspaper by the selected markets. Both the regulated markets transmitted the information to AGMARKNET, Meghalaya State Agricultural Marketing Board (Portal), farmers, Department of agricultural marketing, Department of agriculture and newspaper.
- **Pattern of utilization of AMI and its' benefits :** The extent of utilization of agricultural market information by different categories of sample farmers were in decision making on production, selling and post harvest handling. It was observed that small categories of farmers were not utilizing the AMI on arrivals in decision making on various aspects of production, selling and post harvest handling. Large farmers were used the information on prices in local markets for deciding the crops to be sown, where to sell, when to sell and storage decision. It was observed that small farmers were obtained higher prices (benefits) by utilizing AMI on changing place and time of sale.The extent of utilization of agricultural market information by traders was for decision making on purchase, storage, selling and post harvest handling. It was clearly seen that the agricultural market information was utilized by traders in deciding price to be quoted, quantity to be purchased and the quantity to be store. The traders were also utilized AMI for making decision on when to sale, when to store, quantity to be sold in the study area. It was observed that traders were most benefited by changing time of sale, by mode of storage and change of place of sale.
- **Constraints of AMI faced by Stakeholders:**It was observed that small farmers opined that they had faced difficulty in accessing and not available in time of the AMI. It was clearly seen that non availability of required information on prices, arrivals, area and production was a constraint as expressed by medium farmers and large farmers. Traders expressed that AMI was not available in required form. Traders were also faced difficulty on non-availability of required information on price, prices prevailing other nearby markets, arrivals, area and production aspects. Unable to evaluate and document the agricultural market information by the farmers due to illiteracy and poor communication ability, lack of auction system, no grading from the farmers point and cess was taken in random were the major

constraints in management of AMI in the selected Garobadha and Mawiong regulated markets.

- **Expectations of AMI by Stakeholders:** It revealed that the market information on prices prevailed in other nearby market placed high expectations among all the categories of farmers followed by future price projections and quality wise price information. It was observed that there was an expectation from post harvest handling information for better prices among the different categories of sample farmers. The expectation aspects of traders on AMI indicated that the prices in other nearby markets, future price projections and quality wise prices were given more priority by traders in the study area.
- **Analysis of AMI with Socio-Economic Variables:** It is important to analysesthe relationship between AMI and different socio-economic variables like farm size, caste, type of family, total farm income, education and family size. From the analysis it indicated that farm size, caste, total farm income and education variables were positively correlated with AMI of the sample farmers. Type of family (nuclear or joint), family size were negatively correlated to the AMI access of farmers. It was observed that the socio-economic variables like higher education, total farm income and caste were positively affecting the access of AMI. Joint family system had significant negative effect farmers' access to AMI.

Based upon the results and findings of the case study, the following conclusions and policy implications can be suggested for improving the agricultural marketing information system (AMIS) in East Khasi Hills and West Garo Hills of Meghalaya.

- Empowering farmers with relevant, accurate and timely information about prices being quoted in the market place can help the farmer to take appropriate production related decisions as well as strengthening his bargaining power.
- Of course, sparse inhabitation and geographical barriers worked as a limiting factor in creating desirable agricultural marketing information system (AMIS) infrastructure in remote hills region of East Khasi Hills and West Garo Hills of Meghalaya, but the modern Information and Communication Technology (ICT) in connection with Agricultural Marketing Information (AMI) needs to be given due consideration.
- It is essential to create the pre-requisite for AMIS infrastructure in Garobadha and Mawiong regulated markets of Meghalaya for effective dissemination of information.

- There is need to develop strategies to improve efficiency in data collection, processing, information dissemination and maintenance of databases in the selected regulated markets.
- Training provided to the marketing personnel was inadequate and there is a need to expose the staff to the advance technologies in data management.
- It is necessary toensured flow of regular and reliable data to producers, traders and consumers to derive maximum benefit of their sales and purchases.
- AMIS model should be viable, location specific, user friendly and sustained for a longer period.
- Emphasis should be given on delivery mechanism of information, so that market information reaches timely to the end users in the hilly regions of Meghalaya.
- There is a need to revitalize of APMCs and develop a system of market information utilizing the modern information communication techniques, so that the farmers are provided with the required market information to make appropriate decisions with respect to production and marketing plans including post harvest management storage, processing and sale of agriculture commodities.
- AMIS should create a base for agricultural production planning and marketing led agricultural extension.
- AMIS should be given priority for agricultural marketing strategies and reducing the distress sale at the farmer level.
- Proper integration of various agencies for adequate and efficient dissemination of vital agricultural marketing information, so that it will act as an 'one stop solution' for the needs of the farming community in hilly regions of Meghalaya.
- There is a need to revitalizing the Market Intelligence System especially on dissemination aspects in public institutions like State Department of Agricultural Marketing, Agricultural Universities etc. with modern communication technology.
- The AMI should be deliver fast, reliable and accurate information in a user friendly manner for utilization by the farmers and other stakeholders in order to facilitate the farmers to decide what and when make crop and marketing planning, how to cultivate, when and how to harvest, what post harvest management practices to follow, when, where, how to sell etc. of the agricultural produce in the study area.

- Creating awareness among farmers and other intended beneficiaries on the importance of agricultural market information and its optimum utilization for overall development of agriculture in the state.
- The finding of the present study is guiding the policy makers, administrators, economists, researchers and extension workers to develop, such policies and programmes which would help to increase the efficiency in marketing by effecting improvement in the existing agricultural market information system.

Case Study-3

Supply chain and Marketing Strategy of Pineapple Project

Pineapple production and marketing provides ample scope for farmers to increase their income and is helpful in sustaining large number of agro-based industries which generate huge employment opportunities. It is one of the alternatives for the diversification of agriculture and in fostering and sustaining the tempo of rural development. Effective marketing strategy especially for such a commodity depends mainly on the decision of where, when, how and how much to market. For this the service of a chain of middlemen and functionaries become inevitable. Each of the functionaries and services has to be paid. The marketing system is seen as efficient if the movement of goods from producers to consumers is undertaken at the lowest cost consistent with the provision of services and facilities that consumers desired and are able to pay for. It requires a holistic, multidisciplinary, multilevel systematic and coordinated approach in the identification of and analysis of constraints in the production and marketing. The technological led growth of agriculture in India is recognized as the corner stone of our national strategy for food security, conservation of natural resources, rural employment and development and poverty alleviation. The current national agricultural strategy aims to attain a growth rate of 4 percent plus per annum in the agricultural sector; growth that is based on efficient use of resources and conservation of our soil, water and biodiversity; growth with equity, growth that is demand driven and stabilizes domestic markets and maximizes benefits from exports of agricultural products in the face of global challenges.

Meghalaya in the North-East region is endowed with diverse agro-climatic conditions, rich genetic diversities, vast hydrological resources and pollution free environment that offer a great scope to develop agro-ecosystem a specific technological intervention for diversification of the Agriculture and allied activities viz. horticulture, animal husbandry and fisheries etc. The state displays a distinct ethnic, socio-cultural and geographical identity. Hence, in order to achieve the farming sector as a commercial and profitable enterprise, there is need to develop farm as well as marketing technologies which are simple, cost effective, location specific, eco friendly, economically viable and socially acceptable by the all farming community.

Objectives of the study

i) To find out different marketing channel, price spread and marketing efficiency of pineapple in West Garo Hills of Meghalaya

ii) To examine the disposal trend and prices of pineapple in different categories of sample farms

iii) To estimate the extent of post harvest losses during marketing faced by pineapple growers in the study area.

iv) To identify marketing constraints and suggest policy implication for pineapple market in the study area.

In this case study an attempt has been made to find out (i) Costs, returns and benefit cost ratios in pineapple cultivation (ii)Disposal trends and prices of pineapple (marketing strategies) (iii) Extent of post-harvest losses (iv) Constraints in production and marketing (v) Price spread, Marketing cost, Marketing margin, Producers share in consumers' rupees and Marketing efficiency in different marketing channels in the study area. The present study was undertaken with the sample of 60 pineapple growers of Dadenggre CD Block, West Garo Hills, Meghalaya through stratified random sampling method. The data was collected from the sample pineapple growers through personal interview method with the help of a specially designed schedule covering all aspects. Besides, a sample of 10 wholesalers, 10 traders, 15 retailers and 15 village beoparies in the major marketing centres namely Chibenang and Phulbari Ghat of Selsella CD Block of West Garo Hills were randomly chosen.

From the study,it revealed that the total cost of cultivation was the highest on small farms. It may be due to use of more inputs and higher expenses on labour, planting materials by pineapple growers. It was observed that increase in farm size is accompanied by higher productivity and remunerative price fetched by large farmers as compared to other categories of pineapple growers. The net price received by the all sample farmers was the highest when they sold their produce directly to consumers. The Total costs of cultivation on small, medium, and large farmers were Rs.10,076/-, Rs.7,395/- and Rs.4,965/- per hectare respectively. It was observed that the gross returns by small, medium, large and overall farmers were Rs.37,645/-, Rs.41,130/-, Rs.41,082/- and Rs.39,952/- per hectare respectively. The benefit cost ratio (BCR) of small, medium and large farmers were 3.99, 5.79 and 8.35 respectively. The highest benefit cost ratio was achieved by the large farms because of judicious expenditure in pineapple production and obtaining a sizeable amount of returns.It was evinced that the maximum extent losses of pineapple occur during the time of harvesting at the farm level.The extent of total post harvest losses (q/ha) were 1.02, 1.10 and 0.88 quintal per hectare at the farm level on small, medium and the large farms respectively. It was evident that the total marketing cost of channel-I, channel-II, channel-III, was Rs.105/-, Rs.294/-, and Rs.400/- respectively. The value of goods (pineapple) sold in the channel-I was Rs.700/-, in channel-II it was Rs.1220/- and in case of channel-III it was Rs.1505/-. The indices of marketing efficiency of 5.66 in channel-I was the highest as compared to rest of the channels due to existence of only one middleman. But in case of channel-II and channel-III the efficiency index was 3.14 and 2.76 respectively.

Table 1. Costs, Returns and Benefit Cost Ratios (BCR) in Pineapple Cultivation of Different Categories of Farmers

Particulars	Small	Medium	Large	Overall
Area (ha)	0.87	1.91	3.42	2.06
Production (q)	50.98	118.10	216.15	128.41
Productivity (q/ha)	58.30	61.65	63.27	61.07
Total cost of cultivation (Rs/ha)	10076	7395	4965	7478
Price (Rs/q)	644	660	650	651
Gross Return (Rs/ha)	37645	41130	41082	39952
Net Return (Rs/q)	27499	33735	36120	32451
BCR	3.99	5.79	8.35	6.05

Table 2. Marketing costs, margins and producer's share in consumer's rupees in the marketing of Pineapple through different channels (%)

	Marketing channels		
Particulars	I	II	III
Producer's share	60.71	50.00	44.85
Marketing costs	14.99	24.08	26.55
Marketing margins	24.28	25.89	28.56
Consumer's price	100.00	100.00	100.00

Table 3. Indices of marketing efficiency in different marketing channels

Items	Channel-I	Channel-II	Channel- III
Value of goods sold (V)	700	1220	1505
Marketing cost (I)	105	294	400
Index of marketing efficiency (E)	5.66	3.14	2.76

New paradigm and challenges are needed for pineapple growers of Meghalaya in solving the problem like recurrent price fluctuation, high marketing, storage and transportation cost, non-availability of adequate storage facilities, post harvest losses and lack of competitive marketing system. From the study it will be suggested for improving market infrastructure, direct and group marketing, establishment of modern marketing and processing units, market integration and the improvement of the overall efficiency of the marketing system in the study area. Efficient marketing is a pre-requisite in the development process of Meghalaya economy.To make available of reliable information on village level should require pineapple processing and value addition through small scale industries and cooperative marketing. In concluding remarks, there is an urgent need for promoting producer's cooperative and providing adequate short term credit facilities particularly in the rural areas. In order to hedge risk of pineapple production, it is imperative to develop market intelligence services, introduction of support price and insurance scheme in the state. Meghalaya has become very popular in organic farming which gives very good returns from the West Garo Hills region as well as export market of pineapple produce.

Case Study-4

Value chain on Cashewnut Project

It is evident that apart from the economic importance of cashewnut value chain, it has got greater potentiality in generative income and employment at farm level. It was observed that increase in farm size is accompanied by higher productivity and remunerative price fetched by large farmers as compared to other categories of cashewnut growers. The net price received by the all sample farmers was the highest when they sold their produce directly to consumers. The highest benefit cost ratio was achieved by the large farms because of judicious expenditure in cashewnut production and obtaining a sizeable amount of returns. During the production process, the cashewnut growers experienced the problems of high infestation of pest and diseases, high input costs, scarcity of labour and poor quality of planting materials. Lacks of marketing facilities are some of the problems faced by cashewnut farmers in the study area, for which they are not getting remunerative prices. In the West Garo Hills, cashewnut value chain needs application of modern technology and proper management practices for better production and marketing. The study highlighted that the prospect of cashewnut value chain in Meghalaya is bright as the trend of other traditional crop production in the potential areas is quite encouraging for the organic cashew farming. There is enough scope of enhancing organic produce of cashewnut. Direct Marketing of Cashew produce always fetch the good price for the producer but it is not possible for all categories of farmers. Therefore, it suggest for formation of cooperative society of cashew growers for marketing of various inputs as well as outputs of cashewnut produce.

Objectives of the study

1. To study the economic feasibility of on farm processing of cashew kernels in order to facilitate cashew farmers to have a greater share of consumer rupee.
2. To analyse the existing marketing strategies and value chain of cashewnut into different end products with distinct demand to facilitate entrepreneurs to undertake value addition.
3. To suggest schemes and policies for value chain of cashewnut in different end products

The present study was undertaken with the sample of 120Cashewnut growers comprising of 60 small (1-2 ha.), 40 medium (2-4 ha.) and 20 large (> 4 ha.) farmers of their cashewnut cultivated area from eight villages of Selsella and Dadenggiri CD Blocks of West Garo Hills districts of Meghalayathrough stratified random sampling method. The West Garo Hills District in Meghalaya was selected purposively due to concentration of cashewnut growers in these districts as compared to other district. The data were collected from the sample cashewnut growers through personal interview method with the help of a specially designed schedule covering all aspects.Besides, a sample of 10 Village traders, 10 Retailers, 5 Wholesalers, 3 Pre-harvest Contractors and 2 Processors in the each major marketing centre of Tura, Selsella and Dadenggre of West Garo Hills district was randomly selected. Data pertaining to the agricultural year 2016-17was considered with specific objectives. The project covers one of the horticultural crop i.e.cashewnut with value addition in selected locations. The selected district is namely West Garo Hills, which is having required quantum of cashewnut production. Two blocks from the district was selected for the implementation of the research project. The existing production potential can be suitably utilised for value addition and production of value added products of cashewnut.

It was observed that on an average Cashewnut cultivated area (ha) in case of small, medium and large farmers of Dadenggre were 0.67 ha., 2.35 ha, 4.2 ha. respectively. It was seen that average production of raw cashewnut in case of small, medium, and large farmers were 2.43q, 6.05q and 8.7q respectively. In case of SelsellaBlock, the average Cashewnut cultivated area (ha) for small, medium and large farmers were 0.85 ha., 2.4 ha, 4.3 ha. Respectively; where as the average production of raw cashewnut in case of small, medium, and large farmers were 2.68q, 4.15q and 8.2q respectively. The cost concept analysis in cashewnut cultivation shows that the cashewnut orchard is capital intensive in its initial stage and labour intensive later which shows its cost increased gradually from year to year. During the period of first fourth year of cultivation, the Cost C_2 (Comprehensive Cost) decreases from Rs. 96,620.50 per ha to Rs. 15,926.27 per ha in the year of fifth year in West Garo Hills district. But the cost was found to increased gradually from Rs.17137.42 per ha in the sixth year to Rs.19861.01per ha of the tenth year of cashewnut cultivation. The Cost C_2 was worked out to be higher in the first ourth year of cultivation because of establishment cost which is capital intensive, and the later on its cost was gradually increase from year to year due to labour intensive. The yield of Cashewnut was found to increase gradually from 5.72 q/ha in the year of fifth year to 20.78 q/ ha in tenth year of Cashewnut cultivation. The Gross Farm Income (GFI) was found to increasing from Rs.34,321.43 per ha in the fifth year to Rs.1,24,685.20 per ha in the year of tenth year. The net farm income was found Rs.104824.19 per ha in the tenth year of Cashewnut cultivation. The returns excluding family labour was worked out to be positive from fifth year (Rs. 14416.24 per ha) and Farm business income was estimated to be positive from fifth year onwards in West Garo Hills district (Rs. 2623.84 per ha). The farm investment income was worked out to be positive from sixth year (Rs.21821.79) per ha. of cashewnut cultivation. The Net present value (NPV) was estimated to be positive from eighth year (Rs. 17017.48 per ha) onwards. The Benefit cost ratio (BCR) over total cost was found to be greater than one from fifth year (2.15) onward in the study area.it is revealed that cashewnut is profitable from fourth year and afterward it is more or less benefit upto certain period of the crop life span, so its production need to realize among the farmer community of West Garo Hills of Meghalaya.

Table 4. Economic Returns Analysis of Cashewnut in West Garo Hills of Meghalaya (Rs./ha)

Returns	1st -4th Year	5th Year	6th Year	7th Year	8th Year	9th Year	10th Year
Yield (q/ha)	-	5.72	7.68	10.45	14.02	17.15	20.78
Gross Farm Income (GFI)	-	34321.43	46051.76	62672.58	84122.80	102880.27	124685.20
Net Farm Income (GFI-Cost C_2)	-86620.50	18395.16	28914.34	44679.92	65115.63	83279.78	104824.19
Returns including family labour	-86620.50	-68225.34	-39311.0	5368.92	70484.55	153764.33	258588.52
Returns exclu-ding family labour	-15632.65	14416.24	54937.28	111167.70	187700.48	282304.30	398325.18

Farm Business Income (GFI-Cost A_2)	-30232.66	2623.84	47257.70	108568.58	191463.03	293208.06	416885.37
Farm Investment Income (Farm Business Income-Imputed Value of Family labour)	-45865.30	-17281.35	21821.79	76690.51	151994.94	245463.52	360476.51
Net Present Value (NPV)	-	-60777.60	-40900.91	-15008.61	17017.48	52668.74	91900.95
Benefit Cost Ratio (BCR): Output /Input ratio over Total Cost (GFI/Cost C_2)	-	2.15	2.68	3.48	4.42	5.24	6.27

In West Garo Hills of Meghalaya, Grading and packing of raw cashewnut was done by village merchant, wholesaler and processor (Fig.2). The highest cost incurred by the wholesaler (Rs.200/- per qtl) followed by village merchant (Rs.70/- per qtl of cashewnut) and processor (Rs.60/- per qtl). The cost incurred by wholesaler (Rs.100/- per qtl) for cleaning and drying of raw cashewnut in West Garo Hills of Meghalaya. It was observed that cashew producers were usually incurred cost of value addition of raw cashewnut trough cleaning and sun drying and made disposal of raw cashewnut to the village merchant and wholesaler in the study area. It was observed that the price spread was the highest in channel-III (Rs.3165/-) per quintal of cashewnut in Tura market and the lowest in channel-I (Rs.1040/-) in Dadenggre market. But the marketing of cashewnut through channel-II was the most preferred by the cashew growers. Hence study suggests strengthening the channel-II and encouraging the cashew farmers for better marketing and value addition of cashewnut in the state of Meghalaya. The indices of market efficiency of 11.54 in channel-I was the highest in Tura market as compared to the rest of the channels due to the existence of only one middleman. Similarly the marketing efficiency in channel-I of Dadenggre and Selsella market was 11.29 and 11.26 respectively.

Based upon the results and findings of the study following conclusions and policy implication can be suggested for improving market infrastructure, direct and cashew producer based cooperative marketing and cashew processing units, market integration and the improvement of the value chain of cashewnut in the study area i.e. West Garo Hills in Meghalaya.

- At the primary level, value addition is almost non-persisting. Among other factors, lack of suitable preservation methods at the farmer's end further aggravate the losses.
- Proper integration is required among different cashew growers and agencies who are engaged in production, marketing and processing (value addition) of cashew kernels for designing their future policies.
- Cashewnut in Garo Hills region of Meghalaya is found to be economic feasible which can be more beneficial to reducing the costs of production through intervention of modern techniques in cultivation of the crop.
- Direct Marketing of Cashew produce always fetch the good price for the producer but it is not possible for all categories of farmers. Therefore, it suggest for formation of cooperative society of cashew growers for marketing of various inputs as well as outputs of the member farmers.
- The farmers of Garo hills region are resource poor and unable to afford the input costs as well as expenditures for marketing of cashew produces. Therefore, the financial support is necessary for the cashew farmers through inclusive credit facilities.
- Cost effective agro-advisory services were need of cashew farmers at right time in right form to save the losses.
- Strengthening of most popular channel to enhance the operational marketing efficiency through intervention of post harvest technologies like grading, sorting , packing etc. which will enhance the due share of cashew producer in consumers price.
- The price fluctuations of cashew raw nut have been a major marketing problem as reported by cashew growers, it restrict them for remunerative price of cashew especially during the time of harvesting. Therefore, price forecasting mechanism and expansion of e-NAM in West Garo Hills region is necessary.

Case Study-5

Mushroom Production and Resource Use Efficiency Project

Mushroom enterprise plays predominant role towards employment generation, food and nutritional security and enhancing income. Several small and medium and large companies have taken up mushroom growing on industrial scale to supply good quality of fresh, dried and canned mushroom to the domestic and international markets. Mushroom production is a viable, technically feasible and profitable venture. Since mushroom cultivation is capital-intensive and the financial assistance through institutional agencies at cheaper interest rate would help increase mushroom production. Mushroom cultivation provides additional income to farmers along with utilization of agricultural wastes and thus has bright future in the state. In addition, the left over bed of mushroom may be used as cattle feed and can be converted into quality manure which can generate additional income.

Objectives of the study

In view of the growing demand and importance of oyster mushroom production for promising enterprise, the study was undertaken with the following specific objectives:

i) To estimate the cost and returns of growing oyster mushroom in different farm-size categories,

ii) To identify various determinants affecting its value mushroom productivity in the study area, and

iii) To suggest policy measures to strengthen mushroom production for entrepreneurial development

To estimate the costs and returns from mushroom-growing and explore the factors affecting its value productivity, a field survey was conducted. A list of mushroom growing farmers was collected from the State Department of Horticulture of Meghalaya. Rongram CD Block of West Garo Hills district was selected purposively which has the highest concentration of mushroom growers. From the selected Rongram block, all the 60 mushroom farmers were selected randomly. Data pertaining to the agricultural year 2017-18. The primary data on non-recurring expenses on pucca/ kachcha sheds, platform, plastic sheets and other equipment used in mushroom growing were collected using a specially designed and pre-tested interview schedule as per the objectives. The data on mushroom production and inputs such as paddy straw, plant protection chemicals, human labour, electricity, etc., were collected from the selected respondents. The selected respondents were classified into three categories on the basis of their number of polybag beds spawned using cumulative cube root frequency method. Thus, there were 30 (50.00%) small, 20 (33.33) medium and 10 (16.67%) large mushroom growers in the sample categories.

From the findings of the study, the total cost shows a decreasing trend with increase in the size of beds spawned due to economies of scale. The scale of economies are apparent for both fixed and operational costs. It is revealed that on overall basis, the total cost estimated to be Rs.467 per 5 units of polybag beds spawned. On overall farms, the operational cost account for 83.94% and fixed cost 16.05% of the total cost. In total cost, the expenditure on labour is the most important component accounting for (26.76%) followed by cost on spawn (25.26%). In fact, this basic input of labour explains the positive association with value productivity per bed spawned. According to the different categories of farm size- the total cost per 5-units of beds is estimated to be Rs.518 on small, Rs.483 on medium and Rs.420 on large mushroom farms with a major share of nearly 83% of the operational costs.

It indicates that the oyster mushroom growing is profitable enterprise in West Garo Hills of Meghalaya. On overall farms, the net returns are estimated Rs.683 per 5 units of poly beds. Both the returns over variable cost as well as net returns from mushroom cultivation are directly related to the farm size. It also reveals that the cost of mushroom production decreases with increase in farm size. Thus, larger farmers appear to using the resources more efficiently as compared to smaller farms. The returns from mushroom cultivation depicted a positive association with farm size; these are Rs.58/kg, Rs.76/kg and Rs.120/kg for small, medium and large mushroom farms respectively. It is worked out in the overall farm size that one rupee invested in mushroom growing, yields about Rs 1.55.The input-output ratio is the highest on large mushroom farms (1.82), followed by medium (1.48) and small (1.35) farms.

Table 5. Economic Analysis of Oyster Mushroom Production on Sample Farms

S.No	Particulars	Oyster Mushroom Growers			
		Small farm Farm	Medium Farm	Large	Overall
1	Oyster Mushroom Yield (kg/5 units)	5.12	4.98	5.02	5.05
2	Average price of mushroom (Rs./kg)	225	232	228	228
3	Gross returns from mushroom (Rs./5 units)	1152	1155	1144	1150
4	Variable cost (Rs./5-units.)	436	405	348	392
5	Fixed cost (Rs./sq. m.)	82	78	72	75
6	Total cost (Rs. /5-units	518	483	420	467
7	Net returns (Rs./5-units.)	634	672	724	683
8	Returns to fixed farm resources (Rs./5-units.)	716	750	796	758
9	Cost of production (Rs./kg)	165	158	145	155
10	Variable cost (Rs./kg)	114	102	98	104
11	Returns per kg (Rs./kg)	58	76	120	85
12	Input-output ratio	1.35	1.48	1.82	1.55

The following can be suggested for multiple benefits, making a win-win proposition as a promising mushroom enterprise. It was observed that increase in farm size is accompanied by higher productivity and remunerative price fetched by large farmers as compared to other categories of mushroom growers. The highest input-output ratio (1.82) was achieved by the large farms because of judicious expenditure in mushroom production and obtaining a sizeable amount of returns. It is revealed that mushroom growing is a labour intensive enterprise and it is the most important constituent of the variable cost, followed by expenses on composting and spawning. The net returns have been found higher on large mushroom farms (Rs. 724), followed by medium (Rs.672) and small farms (Rs. 634). Efforts should be made towards appraising the mushroom growers of appropriate techniques and application of modern inputs for better production of oyster mushroom. Input delivery system should be based on the single window approach which encourages in the study area for timely, easy and convenient availability of inputs to mushroom growers. It is necessary for better marketing and value addition of mushroom, so that farmers can get remunerative price for their product. Mushroom farmers should be educated about the farm-size economies of mushroom production. Mushroom growers should be provided training on proper disinfection of pucca-sheds and preparation of good quality compost. There is the need to disseminate the nutritive value of mushroom to increase its consumption and thereby production in the study area. There is a felt need for a linkage and full co-operation between the state government, financial institutions, researchers, scientists, extension workers and mushroom growers on different advanced technological aspects of better production oyster mushroom, effective marketing which help the farmers to get remunerative prices from this enterprise.